Contributions to Environmental Sciences & Innovative Business Technology

Contributions to Environmental Sciences & Innovative Business Technology (CESIBT) is an interdisciplinary series of peer-reviewed books dedicated to addressing emerging research trends relevant to the interplay between Environmental Sciences, Innovation, and Business Technology in their broadest sense. This series constitutes a comprehensive up-to-date interdisciplinary reference that develops integrated concepts for sustainability and discusses the emerging trends and practices that will define the future of these disciplines.

This series publishes the latest developments and research in the various areas of Environmental Sciences, Innovation, and Business Technology, combined with scientific quality and timeliness. It encompasses the theoretical, practical, and methodological aspects of all branches of these scientific disciplines embedded in the fields of Environmental Sciences, Innovation, and Business Technology.

The series also draws on the best research papers from EuroMid Academy of Business and Technology (EMABT) and other international conferences to foster the creation and development of sustainable solutions for local and international organizations worldwide. By including interdisciplinary contributions, this series introduces innovative tools that can best support and shape both the economical and sustainability agenda for the welfare of all countries, through better use of data, a more effective organization, and global, local, and individual work.The series can also present new case studies in real-world settings offering solid examples of recent innovations and business technology with special consideration for resolving environmental issues in different regions of the world.

The series can be beneficial to researchers, instructors, practitioners, consultants, and industrial experts, in addition to governments from around the world. Published in collaboration with EMABT, the Springer CESIBT series will bring together the latest research that addresses key challenges and issues in the domain of Environmental Sciences & Innovative Business Technology for sustainable development.

Naeem Hayat • Noorshella Che Nawi
Editors

Contemporary Issues in Entrepreneurship and Innovative Technology

Springer

Editors

Naeem Hayat
Global Entrepreneurship Research and Innovation Centre
University Malaysia Kelatan
Kota Baharu, Kelantan, Malaysia

Noorshella Che Nawi
Malaysian Graduate School of Entrepreneurship and Business (MGSEB), ANGKASA-UMK Research Academy (AURA), Faculty of Entrepreneurship and Business (FEB), University Malaysia Kelatan, Kota Baharu, Kelantan, Malaysia

ISSN 2731-8303 ISSN 2731-8311 (electronic)
Contributions to Environmental Sciences & Innovative Business Technology
ISBN 978-3-031-50926-1 ISBN 978-3-031-50927-8 (eBook)
https://doi.org/10.1007/978-3-031-50927-8

This Springer imprint is published by the registered company Springer Nature Switzerland AG
The registered company address is: Gewerbestrasse 11, 6330 Cham, Switzerland

Foreword

Entrepreneurship is a beacon of creativity, resilience, and transformation in the dynamic landscape of business and innovation. It is the force that drives economies, fosters innovation, and empowers individuals to shape their destinies. Within this ever-evolving realm, "Contemporary Issues in Entrepreneurship and Innovative Technology" is a vital compass, guiding scholars, practitioners, and students through the intricate terrain of entrepreneurial endeavours in the twenty-first century.

This book is a testament to the collective wisdom and expertise of numerous scholars and practitioners passionate about entrepreneurship. As the editors, Dr. Naeem Hayat and Assoc. Prof. Dr. Noorshella have curated diverse insights, research findings, and practical perspectives illuminating entrepreneurship's multifaceted facets today. Their dedication to advancing the field is evident in the thoughtful selection and organisation of chapters that explore an array of topics, from emerging trends to enduring challenges.

In the following pages, readers will embark on a journey that explores the entrepreneurial ecosystem from various angles. Each chapter is a unique vantage point, offering fresh perspectives on the issues that define entrepreneurship in our contemporary world. The contributors to this volume, who hail from different corners of the globe, bring a rich tapestry of experiences, methodologies, and insights to the discourse.

The entrepreneurship landscape is ever-shifting, influenced by global events, technological revolutions, and cultural transformations. This book endeavours to capture this dynamism by addressing issues that are relevant today and have the potential to shape the future of entrepreneurship. The book's contents span a broad spectrum, reflecting the complexity of entrepreneurship itself. It delves into technological innovation, sustainability, social entrepreneurship, financing, and global business environments, presenting a rich tapestry of ideas, experiences, and perspectives. Through the lens of these chapters, readers will gain a deep understanding of the myriad challenges and opportunities that contemporary entrepreneurs encounter.

Entrepreneurship is pivotal in driving economic growth, fostering innovation, and addressing pressing global challenges in an era of rapid technological advancements, globalisation, and societal shifts. This book is a timely resource for academics, students, policymakers, and practitioners seeking to understand the intricate dynamics of entrepreneurship in our modern world.

As editors, Dr. Naeem Hayat and Assoc. Prof. Dr. Noorshella have meticulously crafted this volume to be a valuable resource for researchers, educators, students, and entrepreneurs alike. Whether you are delving into entrepreneurship for the first time or seeking to deepen your understanding of its intricacies, this book offers a comprehensive and up-to-date exploration of the field.

We invite you to immerse yourself in the following pages and engage with the ideas, research findings, and practical insights presented herein. May this book inspire new perspectives, spark innovative ideas, and contribute to the ongoing dialogue surrounding entrepreneurship as we collectively strive to address the challenges and seize the opportunities of our ever-evolving entrepreneurial landscape.

Dr. Naeem Hayat and Assoc. Prof. Dr. Noorshella have embarked on a journey to illuminate the contemporary issues in entrepreneurship, and we are grateful to them and the esteemed contributors for their dedication to advancing our understanding of this vital field.

Welcome to "Contemporary Issues in Entrepreneurship and Innovative Technology". We hope you find this book both enlightening and inspiring as you navigate the exciting and ever-changing world of entrepreneurship.

Universiti Malaysia Kelantan
Kota Bharu, Kelantan, Malaysia

Noor Azizi Ismail

Preface

Entrepreneurship is a beacon of innovation and resilience in an era marked by rapid technological advancements, evolving market dynamics, and unprecedented global challenges. The Springer Book "Contemporary Issues in Entrepreneurship and Innovative Technology" emerges as a timely and invaluable resource that illuminates the multifaceted landscape of entrepreneurship in the twenty-first century.

Entrepreneurship is a pursuit and a force of transformation for individuals and their societies. This book delves into the heart of this dynamic field, offering an extensive exploration of the challenges, opportunities, and evolving paradigms that shape entrepreneurship today and in the coming days. This volume provides an indispensable roadmap from the seasoned entrepreneur seeking fresh perspectives to the aspiring visionary eager to embark on their journey.

The chapters within this book are not mere theoretical musings; they are the collective wisdom of experts, scholars, and practitioners who have navigated the complex seas of entrepreneurship. They offer insights into a range of contemporary issues, from the intricacies of entrepreneurial finance and sustainable business practices to the role of digital technologies in fostering innovation and growth. Each chapter is a testament to entrepreneurship's vibrant, ever-evolving nature as it adapts to the changing tides of our world. Youth and Women entrepreneurs are the real drivers of creativity, and their efforts are shaping the future of society and business.

As we stand at the crossroads of innovation, sustainability, and global interconnectedness, "Contemporary Issues in Entrepreneurship and Innovative Technology" serves as a compass, guiding us toward a deeper understanding of the challenges and opportunities that define our entrepreneurial landscape. It underscores entrepreneurship's vital role in driving economic development, fostering creativity, and addressing some of society's most pressing issues. Women entrepreneurs are pivotal in endorsing diversity and inclusion that offer a unique perspective to the business world. Women and young entrepreneurs can bring the necessary transformation to build a resilient world future. With the help of technology, these two can contribute to economic growth, job creation, and social progress. The world must support and empower these two to bring the necessary change at the global level.

This book is not a static document but a living testament to the dynamism of entrepreneurship. Its insights are not confined to any one industry or geography; they transcend borders and disciplines. They reflect the universal spirit of human ingenuity and determination, illustrating that entrepreneurship knows no bounds.

This volume offers an understanding toward novel entrepreneurial topics like single mother entrepreneurship, environmental entrepreneurship, green entrepreneurship, micro-entrepreneurship, digital entrepreneurship, cyberpreneruship, and the role of the spouse in women entrepreneurs. The book brings forward the new seas of entrepreneurship and complex grounds nurturing the diverse aspects of entrepreneurial ventures. Women and youth are the two vital facets of the future of entrepreneurship. Groundbreaking solutions can be evolved from the minds of these young and vibrant minds addressing the most pressing global, social, and ecological issues.

We sincerely thank the contributors, reviewers, and the team at Springer for curating this remarkable collection of insights and perspectives. As you embark on this intellectual journey, may you find inspiration, guidance, and the tools you need to navigate the ever-evolving entrepreneurship landscape. "Contemporary Issues in Entrepreneurship and Innovative Technology" is a testament to the enduring spirit of human innovation, and we hope that it will serve as a source of enlightenment and empowerment for all who seek to make their mark on the entrepreneurial stage.

Kota Baharu, Kelantan, Malaysia — Naeem Hayat
Kota Baharu, Kelantan, Malaysia — Norshella Che Nawi

Contents

About the Editors

Naeem Hayat is working as a senior research fellow at the Global Entrepreneurship Research and Innovation Centre (GERIC), University Malaysia Kelantan, Kota Baharu, Malaysia. Dr. Naeem Hayat completed his doctorate in Entrepreneurship in 2021. Before pursuing the Ph.D., Naeem Hayat worked in academia for more than ten years at higher educational institutions in Pakistan. Dr. Naeem completed his post-graduation (MSc HRM) from Napier University, Edinburgh, UK. and MBA (Master in Business Administration) from Islamia University, Pakistan.

Before joining academia, he worked in the consumer banking industry of Pakistan. He worked as a Lecturer at The University of Faisalabad, Faisalabad, from 2010 to 2011. He later joined the GIFT University, Pakistan, as an Assistant Professor of HRM (2011–2018). He is a graduate member of CIPD, UK. His research interests include organizational studies, business administration, entrepreneurship, adoption behavior, and human resources management. His work has been published in Technology in Society, Sustainability, International Journal of Sustainable Agriculture and Informatics, Vine International Journal of Information and Knowledge Management Systems, International Journal of Advanced Science and Technology, and International Journal of Healthcare Management. He published 70 plus research articles and book chapters as well. He is also working in editorial boards in many international journals. He is passionate about teaching and learning.

He conducts training in research writing, data analysis, case writing, and teaching. Naeem delivers research talks and training in Pakistan, Malaysia, China, and Bangladesh. He also worked as a book editor for Springer Books, guest editor for the Sustainability (MDPI) special issue, and editorial board member in a web of science journals. Naeem is also the managing editor for the International Journal of Entrepreneurship Case Studies (IJECS). He also secured international funding and managed projects worth more than USD 10,000.

Noorshella Che Nawi is an Associate Professor at the Faculty of Entrepreneurship and Business at Universiti Malaysia Kelantan (UMK) and Deputy Dean (Academic) at the Malaysian Graduate School of Entrepreneurship and Business (MGSEB), UMK. She has over 12 years of experience in teaching and learning at the undergraduate and postgraduate levels. She holds a degree in Bachelor's Degree in Information Technology (Industrial Computing) (Honors) and a Master of Science (Quality and Productivity Improvement) from the same institution, Universiti Kebangsaan Malaysia (UKM), Malaysia. Meanwhile, her doctorate level was obtained from Victoria University, Melbourne, Australia. She has made noteworthy contributions to SCOPUS and Web of Science-indexed publications. Additionally, she has been awarded more than thirty international and national research grants for various entrepreneurship projects. She leads two women entrepreneurship projects funded by the European Commission and the Korean Government.

List of Figures

List of Tables

Complementarity and Synergies Between SMEs and SEZs Under CPEC: A Mechanism to Revitalize Industrialization in a Developing Economy of Pakistan

Ahsan Abbas and **Muqarab Abbas**

Abstract The small and medium enterprise (SME) sector is considered the backbone of an economy. The sustainable growth of the SME sector is crucial for industrial development, employment creation, import substitution, export promotion, and the economic uplift of any country, especially developing economies. However, due to resource constraints in a developing economy, it is difficult to industrialize to a large extent. For this purpose, developing special economic zones (SEZs) in a country's selected areas is considered the second-best option to efficiently boost investment and effectively develop the industrial/manufacturing sector. Against the backdrop, this study attempts to probe the possible synergies between prioritized special economic zones (pSEZs) under the China-Pakistan Economic Corridor (CPEC) and SMEs through backward linkages, resource mobilization, and knowledge spillover to uplift the country's domestic economy (dominated by SMEs and SEZs). The study modifies and employs the Warr's enclave approach to depict the possible competing and complementing effects of pSEZs on SMEs and concludes that a balanced approach to developing synergy between pSEZs and SMEs is required to increase local investment in the domestic economy as well as attract foreign capital inflows/foreign direct investment (FDI) in pSEZs to complement the SME sector and revitalize industrialization.

Keywords SMEs · pSEZs · CPEC · Import substitution · Export promotion · Industrial development

A. Abbas (✉)
Department of Government and Public Policy (GPP), National Defence University (NDU), Islamabad, Pakistan
e-mail: ahsanabbas@ndu.edu.pk

M. Abbas
School of Economics, International Islamic University, Islamabad, Pakistan

N. Hayat, N. C. Nawi (eds.), *Contemporary Issues in Entrepreneurship and Innovative Technology*, Contributions to Environmental Sciences & Innovative Business Technology, https://doi.org/10.1007/978-3-031-50927-8_1

1 Introduction

In most countries, small and medium enterprises (SMEs) account for more than 90% of all enterprises and contribute significantly to economic growth. Countries around the globe emphasize this sector's role in stimulating economic growth (Zafar & Mustafa, 2017). Given the important role of SMEs in the economy, countries of the world coordinate with each other to develop a comprehensive SME expansion framework that clearly defines the country's role and responsibility for providing comprehensive support to develop SME sectort (Rohra & Panhwar, 2009).

The small and medium enterprise sector is the backbone of an economy as it contributes to export growth, employment creation, and poverty alleviation of the economy. The domestic economy of Pakistan is mostly SME based; i.e., 90% of all 3.2 million enterprises in Pakistan are SMEs. SMEs employ 70% of the nonagricultural labor force and contribute about 40% to the country's gross domestic product (GDP). Moreover, their share in value addition is around 35%, and they add 25% to export earnings (Aslam et al., 2018).

It is pertinent to note that the industrial/manufacturing sector, especially the SME sector, faces manifold challenges, like scarcity of working capital, lack of educated and skilled manpower, low factor productivity, managerial inefficiencies, entrepreneurship void, high input cost, obsolete technologies, and limited access to new technologies and novel practices. Additionally, a lack of proper branding, marketing strategy, packaging, and effective managerial skills is impeding the decent growth of the sector (Dar et al., 2017). To overcome the issues faced by the SME sector, the government has taken many initiatives, and many institutions, like the Small and Medium Enterprise Development Authority (SMEDA), the National Vocational and Technical Training Commission (NAVTTC), and the Technical Education and Vocational Training Authority (TEVTA), have been set up for the development of SMEs in Pakistan. Especially, the Small and Medium Enterprise Development Authority (SMEDA) has been established by the Government of Pakistan to facilitate small businesses. SMEDA is working on registering SMEs, profiling clusters, providing marketing support, and training industrial clusters for skill development through the National Vocational and Technical Training Commission (NAVTTC) and the Technical Education and Vocational Training Authority (TEVTA).

Although NAVTTC and TVETA have introduced various technical and vocational training courses to provide a skilled and trained labor force to the industrial sector, there still exists inefficient labor, with low skills and productivity. Furthermore, unskilled labor, running the family business, and a lack of cooperative culture are important reasons for industrial backwardness (Kemal, 2006; Sabir & Sabir, 2010). Unfortunately, in Pakistan, infrastructures like roads, transport, railways, gas, electricity, and communication are not developed and accessible as per the requirements of the industry. In addition to this, the costs of energy and key inputs are also very high compared to those of regional countries (Mahmood et al., 2014). Similarly, the lack of capital is a major problem in industrial development; to resolve it, industrial estates are needed as a source of foreign investment (Balasubramanyam, 1988). Pakistan faces a deficiency of capital and a low saving rate; i.e., the gross saving rate was 5.7% of GDP (World Bank, 2021). Therefore, the lack of capital is a strong hurdle in industrial development. Furthermore, the deficient capital poses a significant challenge toward technological development i.e., inhibiting innovation, and hindering firm's ability to become productive and competitive.

Moreover, to remove infrastructure bottlenecks, the government has developed several industrial estates/parks and economic zones, mainly for import substitution. For export promotion purposes, export processing zones have, on the other hand, been established. The infrastructure of these industrial estates or parks plays a vital role in removing the basic problems faced by industries, like transportation, shortage of energy (electricity or gas), and lack of capital. Rietveld (1989) identified the three approaches to solving industrial issues namely, 1) Industrial estate infrastructure as itself a model as a production factor in the production function. 2), Industrial estate infrastructure is mainly a location factor that influences public and private investment and employment in a dedicated manner. 3) The importance of industrial estate infrastructure can be analyzed via its impact on international trade. These industrial estates are the hope of developing countries as their products will be exported to foreign markets and will be the source of foreign earnings (Ali & Faisal, 2016).

In Pakistan, the industrial structure itself is a big issue. The production of domestic consumer goods is preferred over the export of foreign capital goods. The industrial production structure depends on self-interests, not on the country's requirements. There is no doubt that there is an additional profit in consumer goods, but capital goods are required for a country's development. Similarly, there are limited markets for Pakistan's industrial goods abroad. The lack of export markets is a major obstacle to industrial development.

Moreover, the increase in GDP depends on the growth of the industrial sector, which has two major subsectors: small scale and large scale. Figure 1 depicts the

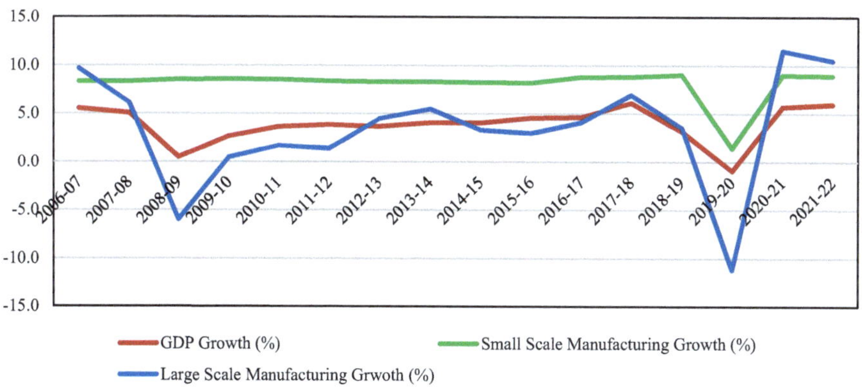

Fig. 1 GDP and manufacturing sector growth rates. Source: *Economic Survey of Pakistan 2021–22*

growth of industry and the GDP of Pakistan. Over the years, the large-scale manufacturing sector has experienced fluctuations, while the small-scale sector has had a smooth growth of around 9% per year until the emergence of COVID-19.

Figure 1 indicates that the large-scale sector is volatile, while the small-scale sector has a somehow constant growth pattern. The growth pattern of the large-scale sector makes it uncertain, which leads to low investment as investors are cautious about investing in an uncertain sector (Liu et al., 2023). Although, the small-scale sector is an impetus for the industry but its share in the overall GDP of the country is meager which indicates that the sector is completely ignored as the sector faces unfavorable policies (Haque, 2007). Figure 2 indicates the share of the small- and large-scale sectors in the GDP of the economy.

Although the small-scale sector has lower standards, fewer finances, and a poor organizational structure, it has global importance; it creates employment opportunities and social cohesion and plays an important role, especially in developing countries (Du et al., 2023; Hyder & Lussier, 2016). This is because small-scale enterprises are flexible in terms of making new decisions, adopting updated technologies, and switching production patterns (Song et al., 2023). It is pertinent to note that while the small-scale sector has not satisfactorily progressed, its contribution to GDP has marginally increased from 1.53% to 1.98% over the last 10 years.

Additionally, the main causes of backwardness are political instability and economic uncertainty. The frequent changes in policies and management and inconsistent policies can be considered key causes of deindustrialization in Pakistan. Most industrial owners are political leaders or have political relations, so they can frequently change the policies for their interests. Thus, foreign investors feel they are making a risk due to uncertain political situations and interests (Goncharuk, 2007).

To overcome the above-cited problems in industrial growth and development, Pakistan has recently taken modern, integrated, and technically efficient steps to provide basic industrial infrastructure, like an effective one-window operation (one-stop shop), the development of land, and the provision of uninterrupted power/electricity and gas supply, as well as other requisite amenities. Furthermore, a

Fig. 2 Percentage share in GDP of the manufacturing sector. Source: *Economic Survey of Pakistan 2021–22*

comprehensive road and railway network connecting domestic and foreign markets, mainly through the intervention of the China-Pakistan Economic Corridor (CPEC), will support the development of backward and forward linkages. In this way, the issue of limited market access would be resolved, and it would link Pakistan to the global value chain, enhancing export. In this regard, the establishment of prioritized special economic zones (pSEZs) can be considered one of the key initiatives. The term special economic zones (SEZs) includes an extensive series of zones (see Table A in Appendix), like export processing zones (EPZs), free trade zones (FTZs), industrial parks, free ports, and enterprise zones (Akinci & Crittle, 2008).

Although these zones definitionally/geographically vary from each other, but each one of them has the same characteristics. Broadly, Zeng (2016) deliberated the four features to define the concept of special economic zones (SEZs) these include; (1) a geological-specific area, usually physically protected; (2) the existence of one administration or management; (3) benefits for investors substantially within the zone; (4) separate customs zones (duty free) and simplified procedures. Accordingly, there are seven approved/notified (see Table B in Appendix) SEZs all over Pakistan. Additionally, the government has proposed nine prioritized SEZs (pSEZs) under CPEC to industrialize Pakistan in the medium to long term (see Table C in Appendix).

1.1 Aims and Objectives of the Study

The study attempts to probe the possible linkages and synergies that may exist between pSEZs under CPEC and the domestic economy/SME sector. The analysis examines how backward linkages can be established through the domestic economy of Pakistan, dominated by SMEs with the liberalized economy of the SEZs. Comparative advantage, cost-effectiveness, and export competitiveness can comprehensively be achieved through complementarities and synergies between pSEZs and SMEs. Furthermore, the study also investigates/identifies the competing channels through which pSEZs and SMEs may confront cutthroat competition and result in crowding out effects to marginalize the advantages of undertaken initiatives. Finally, the analysis seeks to figure out whether the SME sector of Pakistan can be integrated into the global value chain via the development of pSEZs and the intervention of CPEC so that a win-win situation can be created.

1.2 Motivation and Significance

The manufacturing sector, especially the SME sector, is an essential component of commodity-producing sectors, namely, agriculture and industry. The data reveal that not only the share of this key economic activity is shrinking over time but also its growth shows stagnancy and descending trends. Historically, Pakistan's economy has undergone a process of deindustrialization so, the dream of achieving decent and sustained growth in the developing economy of Pakistan cannot come true without sustainable development of the industrial notably manufacturing sector. No study, so far, has attempted to examine/analyze the inclusive, balanced, and synergized industrial development through harmonizing the old and new economic activities such as complementing existing SMEs with modernized SMEs. The present study, however, attempts to fill this gap.

1.3 Plan of the Study

This study is organized as follows: the next section sheds light on the complementarity between SMEs and SEZs focusing on pSEZs under CPEC. The relevant literature is reviewed in the section two. Section three highlights the theoretical framework and also explains the SEZs experiences of the various countries. The final section concludes the study and proposes way forwards.

1.4 SME-SEZ Complementarity and an Overview of Prioritized SEZs (pSEZs) Under CPEC

Special economic zones (SEZs) are developed for the industrialization of an economy as well as for the acceleration of economic growth. It is considered the "second best" and cornerstone strategy for economic growth as it is not possible to industrialize an entire country at once due to the scarcity of resources (Farole & Akinci, 2011). SEZs offer many benefits, such as attracting foreign direct investment (FDI), generating employment, reducing poverty, speeding up industrialization, increasing foreign exchange, human capital formation and development, technological advancement, and knowledge learning and sharing (Edge & Rudd, 2011). SEZs are considered a strong economic instrument for fueling the industry, backed by quality infrastructure, regional connectivity, uninterrupted power supply, facilitation services, and fiscal benefits, which in turn will boost the economy.

SEZs attract foreign investment in Pakistan, which can resolve the problem of limited capital for industry (Farole & Akinci, 2011). Foreign investment brings modern technology also, which increases the productivity of existing labor and capital (Borensztein et al., 1998). Foreign investment in SEZs can create new jobs, which absorbs labor and decreases the unemployment rate in Pakistan. The industry design in SEZs is export promotion instead of import substitution, which gets foreign earnings, boosting the industrial sector as well as the overall economy (Lin & Monga, 2011).

A country needs sustained industrial development. Accordingly, not only establishing and bringing new industries mainly in special economic zones (SEZs)/export processing zones/free trade zones is important but also sustaining existing industries and SMEs. The incentives offered to local and foreign investors and the possible relocation of efficient foreign companies and firms in SEZs are the major concerns of existing SMEs. SMEs face financial obstacles, managerial issues, training problems, and infrastructure issues.

The SME sector in Pakistan is not reasonably progressing and significantly growing. To kick-start industrialization in Pakistan, along with seven approved/notified non-CPEC SEZS, nine prioritized SEZs are proposed under CPEC. It is planned to develop, on a fast-track basis, three pilot prioritized SEZs (pSEZs), namely, the Rashakai Special Economic Zone, Nowshera; Allama Iqbal Industrial City, Faisalabad; and the Dhabeji Industrial Zone, Thatta. In this regard, existing SMEs can be linked with related industries in the above-mentioned pilot pSEZs. So far, none of these pSEZs is in their working stage, so empirical analysis is not possible yet. While in the light of existing studies (Lu et. al., 2023; Arbolino et. al., 2023), it is considered that SEZs are a kick start with SMEs to industrial development and economic growth. Additionally, there is a possibility of backward linkages between SMEs and pSEZs for the case of Pakistan's industrial and economic development. Furthermore, the possible relocation of export oriented Chinese/foreign companies can develop forward linkages and through technology transfer/capital inflow expand the international market for made in/by Pakistan products and link with the global value chain (GVC) of SEZs with SMEs.

Essentially, the industrialization process is expected to take off in pSEZs under CPEC, which is a huge opportunity for Pakistan, to strengthen its weak industrial base/structure and to flourish already existing industries through establishing backward linkages and business alliances/joint ventures, especially with Chinese companies. Local SMEs are lagging behind because of underdeveloped infrastructure; the lack of finance, capital equipment, and expertise; and insufficient energy availability. These issues are being addressed under various CPEC projects, especially the development of SEZs in all the provinces of Pakistan. Therefore, complementarity and synergy between SMEs and SEZs will pave the industrial process in the Pakistan economy. Further, it will result in export expansion, product promotion, and import substitution through enhanced capacity of local production to develop high-quality and value-added goods.

Light engineering products, namely, low-tech machinery, power loom spare parts, knitting machines, winder machines, warping machines, sizing machines, interlocking machines, wheat threshers, choppers, plowing blades, etc., are being manufactured locally. Despite having experience in producing such machines, however, Pakistan is still lagging behind in the production of high-technology capital equipment. Sophisticated industrial machines are being produced at a small scale, though they have the potential to be fully produced in the country.

Based on indigenous endowments and latent comparative advantages (natural resources, demographic dividend, accessibility, connectivity, diverse economic activities, regional demand, prevailing international trade scenario, and geo-economics) the industrial potentials in the above-mentioned fast track pSEZs under CPEC are highlighted below. The sectors/clusters in each pSEZs for both import substitution and export promotion are figured out on the basis of backward linkages:

human resource, raw material/semi-finished material/intermediate goods, logistics connectivity, and possible forward linkage i.e., existence of local and regional consumption.

1.4.1 Industrial Potential

Rashakai Economic Zone (REZ)

The following industrial clusters/sectors are proposed/planned:

- Wooden furniture
- Marble, precious stones
- Tobacco, sugar
- Garment and textile products
- Home building materials
- General merchandise
- Electric machinery, equipment, and parts
- Electronics and electrical appliances
- Automobile and mechanical equipment
- Culinary
- Fish, food sector

Dhabeji Special Economic Zone (DSEZ)

DSEZ has the potential to transform into an expatriate enclave with modern infrastructure and a fiscal incentive package. Sindh Economic Zones Management Company (SEZMC) of the Government of Sindh is the provincial agency for the development and management of the Dhabeji SEZ. The following industrial clusters/sectors are proposed/planned to be established in DSEZ:

- Steel foundries
- Automotive and auto parts (rubber tires, etc.)
- Silicon raw, chemicals and pharmaceuticals
- Metal-derived industries
- Renewable-energy-related items
- Textile and garments
- Warehousing
- Building materials
- Consumer packaged goods (CPG)/fast-moving consumer goods (FMCG)

Allama Iqbal Industrial City (AIIC)

The Allama Iqbal SEZ has the advantage of being adjacent to an M3-Industrial city, which comprises many projects, including textile, pharmaceuticals, information technology, chemicals, automotive, service complex, etc. The proposed industrial sectors/clusters for AIIC SEZ, Faisalabad, include the following:

- Textile and apparel, textile value added
- Footwear
- Electric fittings and toys
- Plastic-derived products
- Appliances
- Leather
- Paper and pulp
- Mechanical machinery
- Culinary
- Food and fish products

Integration of existing SMEs with SEZs/pSEZs is a big challenge, concerns of the local industry, and the apprehensions of the former to compete with efficient and competitive firms in the enclaves i.e. SEZs/pSEZs. The SEZs have diverse effects on labor/capital markets for example, when labor/capital demand increases in the SEZs, it leads to an increase in the overall demand for labor/capital and cause upward pressure on factor prices (House & Shapiro, 2006). It will drastically effect the competitiveness of the existing SMEs across the economic activities (Cirera et al. 2014). Keeping in view the situation of low productive and unskilled labor force of Pakistan (Khan & Khan, 2010), any further cutthroat competition and substitutability between SMEs and SEZs will emerge as a disaster for the industrialization process and economic development.

2 Literature Review

There is enough evidence in modern literature that establishes a strong link between knowledge spillover and the structure of spatial economies that contribute to regional economic development (Fingleton & López-Bazo, 2006). Successful special economic zones (SEZs) can attract a large number of international and national companies, promote economic development, and make significant contributions to business investment, job creation, and employment. It is estimated that in China, special economic zones create 30 million jobs, accounting for about 22% of the country's GDP, 46% of foreign direct investment, and 60% of exports (Zeng, 2016).

The special economic zones (SEZs) can efficiently and effectively accomplish industrial development through efficient utilization of limited funds and cluster expansion. Investing in SEZs provides hustling community amenities in a geographically concentrated area and promote urban development through the provision of high-skilled workers and requisite facilities, beneficial to improve living standard (Lin & Monga, 2011).

Likewise, Johansson and Nilsson (1997) find that SEZs have encouraging influences on exports in Asia, particularly Malaysia and Sri Lanka. Their results specify that SEZs are more likely to have a positive impact on exports when countries adopt outward-looking, export-oriented policies at the national level to promote their SEZs. In Costa Rica, from 1990 to 2003, export share jumped from 10% to 55%, and export items changing from textile products to modular circuits and other electrical products is the successful result of SEZs (FIAS, 2008).

Essentially, special economic zones attract a proper mix of foreign investors, who bring knowledge of how to master the manufacture, promotion, and distribution of export commodities. This familiarity translates into indirect impacts on local markets, stimulates local enterprises to enter export markets, and increases the production of export products (Johansson & Nilsson, 1997). Similarly, Rodriguez (1976) suggested the opportunity for capital mobility between SEZs and other sectors of the economy and also concluded that the economy will attain the same level of welfare as can be achieved under an overall free trade regime.

In contrast, Aggarwal et al. (2008) stated that the role of SEZs in export diversification has varied by country across sectors and products. SEZs required geographically suitable locations and specific and desired production decisions without political interference. Similarly, in the absence of necessary ingredients and decisions, SEZs are likely to be a burden for the whole economy rather than an instrument for development and growth (Balasubramanyam, 1988).

While SEZs manufactured final/furnished goods by using intermediate goods as input produced by existing SMEs, thus, in this way SEZs are also a source to encourage links by transfer technology and boost the domestic economy preferably SMEs (Warr, 1989). Similarly, Din (1994) discloses that SEZs support the domestic manufacturing sector through backward linkages, like using domestic manufacturing inputs for further processing. Lall (2000) concluded significant evidence indicating that SEZs have played a vital role in industrial development and the transfer of technology to new industrialized East Asian economies, especially South Korea, Taiwan, Malaysia, and the Philippines. There has been a big industry upgrade to the electronics sector mostly within the zones.

SEZs have direct economic benefits to the economy in terms of job creation and factors productivity enhancement as well as indirect benefits including capital accumulation, foreign exchange earnings, technology transfer and managerial expertise for the host economy (Ali & Faisal, 2016). Meanwhile, Zeng (2016) concluded that the direct impacts of SEZs are diverse across countries and regions. He also studied the broad objective of SEZs and concluded that there is no way to measure the success of SEZs. Rhee et al. (1990) surveyed local and foreign firms as well as their joint ventures functioning within the zones and concluded that foreign firms influence local firms and the export logistics supply chain. An extensive variety of products is made by foreign firms by employing advanced technology that was also adopted by the local firms.

It is found that SMEs face challenges to employ the right management and skills for market based requirements and product upgradation. The technical and productive labor at the right place is a roadmap for sustaining the growth of SMEs (Parida & Madheswaran, 2023). Traditional SMEs are motivated to adopt new technology, learn managerial skills, and employ effective marketing techniques in connection with modern SEZs. These changes lead to the enhancement of the business environment, the encouragement of investment, the creation of labor demands, and an increase in exports (Thamwicha & Chaiprasit, 2018).

However, SEZs are only beneficial for developing economies if they can attract FDI in labor-intensive sectors. Further, there are diverse effects in the case of intermediate goods traded internationally at a competitive price and intermediate goods that are traded domestically at a domestically settled price; the same diverse effects happen in the case of capital inflows (Krugman, 1991; Din, 1994). On the other hand, capital-intensive goods are demanded internationally and may be a case of foreign capital inflows. Furthermore, if intermediate goods/SMEs produce are traded internationally with final goods, it will also earns foreign exchange. So, in this way, SEZs will boost the economy and attain capital/financial inflows along with SMEs (Arndt et al., 2010).

201(2014

During the last decade, SEZs assist the development for low income countries, it is because these zones are the source of pulling investment, increasing trade, and opening opportunities for labor (Jiyamuratov, 2023). While Aggarwal (2023) indicated that the SEZs could not be insulated from the wider institutional contexts in which they are embedded. Between promises and practical implementation, and between perceived and actual improvement, there exist gaps.

Pakistan faces issues like low investment, poor quality of institutions, an inefficient labor force, and a lack of capital (Hussain & Rao, 2020). In this regard, SEZs are a major source of investment, technology inflows, and technical support and can increase labor productivity by providing skills and technical education (Hadi et al., 2018). Additionally, the institutions will be strengthened through foreign investment inflows, contract enforcement, respecting property rights and by adopting the international rules and regulations (Khaskheli et al., 2023).

The literature concluds that SEZs are important for industrial development, employment generation, technological transfer, and overall economic development. But the linkages between SEZs and SMEs are still unattended. In this regard, the purpose of this study is to develop possible complementarity and synergies among SEZs, SMEs and other domestic firms outside the SEZs.

3 Framework of Study/Methodology

The framework devised by Warr (1989) is modified to establish the link between SMEs and SEZs, and flows of physical and financial resources as well as goods and services. The model mainly explains and assesses the inflow of expected foreign capital or investment in the prioritized SEZs (pSEZs) under CPEC which is mainly focusing on possible backward linkages with existing SMEs. Moreover, the role of SEZs in forward linkages and export promotion (global value chains) is also highlighted. Importantly, it is noted that SEZs accrue direct and indirect economic benefits, like employment generation, foreign capital inflow, technology transfer, knowledge spillover, and urban development (Din, 1994; Zeng, 2016).

3.1 *Description of Model and Analysis*

The proposed model schematically establishes the nexus among SMEs/existing industries and SEZs. In light of Warr's description of the enclave approach, the devised model (diagram 1) depicts the possible competing and complementing effects of SEZs on the domestic economy/SMEs. The various flows, shown in the below schematic (flows from A to B and A to C) demonstrate that SEZs and SMEs compete to acquire limited domestic resources. Additionally, the cutthroat competition between efficient and large-scale multinational companies (MNCs) and local SMEs can turn out to be havoc and has a devastating effect on existing domestic industry (Blomström et al., 2001). Domestic labor is demanded by both SMEs and the SEZs, while the large-scale firms in SEZs may offer higher remuneration and job return to the domestic labor as compared to SMEs and may be able to capture higher share and attract more skilled/quality labor because of their efficiency and managerial expertise.

Similarly, tax incentives and subsidies offered to SEZs are relatively in favor of SEZs than existing industries/SMEs mainly operating outside the territory of SEZs. So, in the case of subsidies and tax incentives to SEZs, SMEs will also be unevenly competing. Moreover, firms operating in SEZs and SMEs will confront in the acquisition of raw materials as well. Additionally, the SMEs lack managerial expertise, and the efficient large firms in SEZs will be more capable of hiring better management from the domestic economy, because of their lucrative returns and fringe benefits (Iršová & Havránek, 2013).

SEZs can also compete with SMEs (Blomström et al., 2001) mainly in the selling of commodities (in terms of import substitutions) in the domestic economy. Moreover, the competition between the two entities is also possible in capturing the world market (exporting of goods and services). In this regard, the SEZ sector has a comparative advantage and competitive edge over SMEs because of their relatively low cost of production and economies of scale and scope. To take full benefit from special zones, governments and zone management need to consider local comparative advantages while targeting priority sectors. Governments and zone management should also help local firms link with zone investors through supply chains or subcontracting relations.

Moreover, flows from B to D and also from C to D indicate that both SMEs and SEZs may absorb international managerial expertise, technology transfer, knowledge spillover, and global best practices (see blue arrows in the below diagram). SEZs also purchase capital and intermediate goods (machinery and equipment) from abroad, which results in the outflow of foreign currency and may cause pressure on exchange rates as well as foreign exchange rationing (see black arrows in the following diagram). Similarly, production in SEZs may be directly exported to foreign firms, which may not impact the overall economic growth of the country.

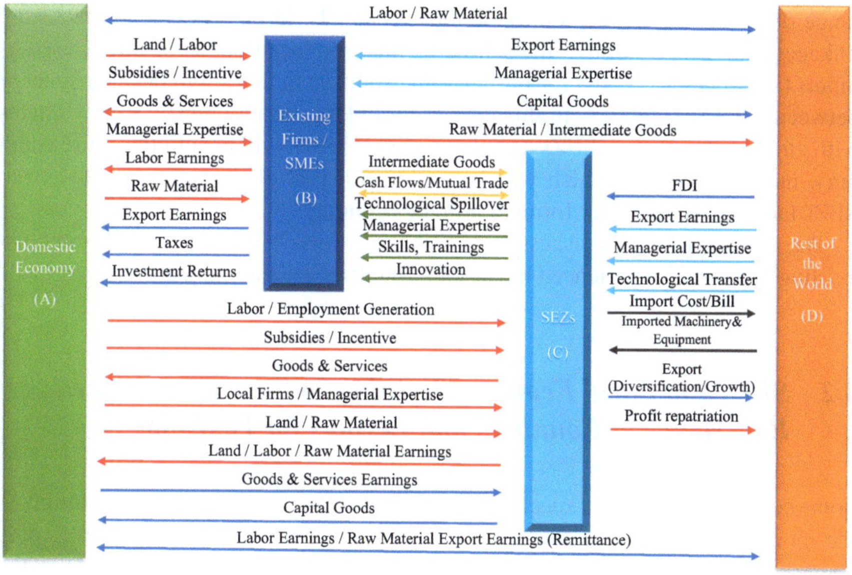

Diagram 1 The enclave model of the nexus between SMEs and SEZs

Revenue collection and government earnings will not be considerably affected by officially declared profit and losses, exemption from profit, sales tax discount, as well as several other taxes, like corporate income tax (Warr, 1989). Foreign-owned companies will be mainly operating in SEZs, which will be using foreign currency to hire domestic labor and raw materials; in this way, the host economy gets foreign exchange in return for labor and raw material earnings. Notably, the key component in the SMEs and SEZs nexus is possible complementarity or substitutability effects channelized through investment flows i.e., foreign (direct) investment and domestic investment in the respective sector.

Similarly, SEZs also create competition (cutthroat) between efficient (large-scale) foreign companies, including MNCs having modern technology, and inefficient (small-scale) domestic companies using obsolete technology. MNCs/foreign firms have extensive access to limited local resources because of sufficient capital and the possible disruption of backward linkages by producing (import substitution) local goods demanded by the domestic consumer. Notably, these foreign firms are more competitive because of high managerial expertise and (fiscal/commercial) incentives offered by the host country to attract FDI.

SEZs in a positive since creating novel investment opportunities, access to technology and equipment, imitation, possible business alliances/joint ventures, relocation and entry of new industries, knowledge spillover effect, technology transfer, HRD, decrease in transaction/operational costs, improved managerial expertise, infrastructural development, and creation of backward linkages.

Finally, terms of the cooperative and complementary relationship between SEZs and SMEs, namely intermediate goods, managerial expertise, technology, and cash flows are shown by green and yellow arrows. Accordingly, SMEs can get a wide range of benefits from SEZs through these channels. In this regard, the backward linkages established through the selling of final produced goods of SMEs used as inputs (intermediate goods) for further value addition in SEZs will develop synergy between these two sectors. In this regard, by linking local/domestic supply chains with global value chains, SEZs can play a vital role that will benefit the entire economy, including SMEs, which will be an integral part of this process. Moreover, SEZs facilitate spur cluster formation in the economy.

The following section discusses the linkages between SEZs and local economy/existing industries experienced by various countries around the world.

3.2 SEZs and Local Economy/Existing Industrial Linkages: Experience of Countries and Lessons for Pakistan

Some economists consider special economic zones as the "second-best solution" to the antiexport bias of trade policies and other policy distortions in many developing countries. Using various hypotheses and parameters, economists concluded that the zones lead to economic distortion in the region (Hamada, 1974), on the contrary, the establishment of a zone has improved the well-being of a country (Young & Miyagiwa, 1987). At the policy level, the basic discussion is whether special economic zones are used as demonstration zones or catalysts to promoting national economic policy reforms, or as "pressure valves" for unemployment, and reducing the incentive to reforms. Yuan and Eden (1992) explored that the developing countries used zones to muddle along without reforms and strained the requirement to use zones as a supplement to countrywide reform, rather than creating isolated free-market enclaves.

In this regard, the experience of the Republic of Korea and the representative of Dominica led to a rich disparity. The Republic of Korea is an example of a country where 500 zone firms purchased not more than 0.01% of the input material from the local firms even after the completion of 30 years of the first industrial zone in Korea, few regions have successfully coexisted with a highly protected domestic economy (Schrank, 2001). Similarly, in the Philippines, owing to the high cost of government infrastructure and subsidized utilities for the development of the Badan Export Processing Zone, the results were negative (Warr, 1989).

According to Schrank (2001), EPZs have failed to become a "bridge of structural reform" in small economies, and they have not been able to become the efficient and competitive in the case of import substitution. In these cases, export processing zones remain isolated enclaves "at the expense of national industry." In larger economies, where an industrial foundation has already been established, backward linkages are more successful.

While different countries, such as China, Malaysia, Kuwait, Jordan, Sri Lanka, and Ireland, have made deliberate efforts to integrate zones into national economies and facilitate a broader export orientation, transfer of technology and improve the overall business environment by extending best practice policy. The zones become economically efficient and enhance increasing returns well above the estimated opportunity costs of the respective countries (Akinci & Crittle, 2008; Kankesu, 2003).

Most of the economic zone initiatives taken by Europe and Central Asia do have at least some success, where Poland, Bulgaria, and Romania are at the forefront. In Ukraine, issues in investment, locations and facilities, as well as public sector management capacity, has undermined operations, and a large proportion of regional investment turns out to be the result of relocation rather than a new effort (Akinci & Crittle, 2008). While, in China, the economic zone is mainly used as a way to implement national and regional development strategies to pave the path of economic development and urbanization (Zeng, 2016).

4 Conclusion and a Way Forward

The sustainable development of SMEs through SEZs is modeled and analyzed in the study. Both the positive and the negative effects of SEZs on SMEs are highlighted through the proposed/devised schematic model (enclave model). The main focus of the study is to evaluate the positive effects/spillovers of SEZs on SMEs (complementarity and synergies between the two entities). Accordingly, backward linkages between SMEs and SEZs as well as prospective economic/business alliances (i.e., joint ventures, networking, licensing and franchising, etc.) in different stages of product development are depicted and discussed. Further, the analysis provides the insights for forward linkages to support and strengthen the manufacturing sector resultantly reverse the deindustrialization in Pakistan.

The government needs to take necessary policy measures mainly through public-private partnership modes in the areas where complementarity and synergies exist. Moreover, the possible joint ventures in potential sectors and product lines may strengthen the positive link between SEZs and SMEs. In this regard, SMEDA (2019) has also proposed the following industries/products that are potentially suitable for joint ventures:

- Light engineering (textile machinery, agricultural machinery, textile spare parts)
- Power looms, auto part manufacturing
- Cutlery manufacturing
- Auto part design and inspection center
- Electrical fitting industry
- Wooden furniture manufacturing
- PVC pipes and plastic products
- Marble and granite firms

Keeping in view the sustainability of SMEs and the success of SEZs, viable backward linkages have great significance and are highly important. For this purpose, pSEZs under CPEC should be mainly focused on export-oriented sectors, where the potential has not been properly exploited so far.

However, balanced and coordinated approach to develop synergies between SEZs and SMEs is required for the industrial development so that the foreign capital inflow/investment in SEZs should complement the SME sector. Finally, demand-driven technical education should be imparted to effectively and efficiently enhance the skills of industrial workers. In this way, the rate of return from foreign financial inflows can outweigh the associated cost of investment, which can bring decent growth in Pakistan's industrial sector. Pakistan should learn lessons from other countries and not repeat its past mistakes.

Appendix

Table A Classification of special economic zones

Type of zone	Development objectives	Description	Markets	Size
Free trade zones (FTZs)	Supporting trade	Also known as commercial free zones, these are clearly delimited areas offering warehousing, storage, and other services aimed at boosting import-export.	Domestic, re-export	<50 hectares
Export processing zones (EPZs)	Export and manufacturing	Industrial clusters offering incentives and the facilitation of manufacturing and other activities that are prevalently export-oriented.	Mostly exports	<100 hectares

(continued)

Table A (continued)

Type of zone	Development objectives	Description	Markets	Size
Hybrid EPZs	Export and manufacturing	Subdivided zones with one part open to all industries regardless of their export orientation and another specifically designed for export-oriented firms.	Export and domestic markets	<100 hectares
Freeport	Integrated development	Large territories accommodating all types of activities, providing broad incentives and benefits; can also include residents on the site.	Domestic, internal, and export markets	>1000 hectares
Enterprise zones	Urban revitalization	Aimed at revitalizing distressed urban and rural areas; enterprise zones mostly flourish in developed countries and are characterized by the provision of tax incentives and financial grants.	Domestic	<50 hectares
Single-factory EPZs	Export manufacturing	Incentives are provided to a specific enterprise rather than to a geographic location.	Export market	No minimum

Source: *World Bank (FIAS) Report*, 2008

Table B Nine prioritized SEZs (pSEZs) under CPEC

S. no.	Name	Location	Potential industries	Area	Connectivity
1	Allama Iqbal Industrial City	Faisalabad, Punjab	Textile, steel, chemicals, pharmaceuticals	3217 acres	Motorway (M-3)
2	ICT model industrial zone	Islamabad	Pharmaceuticals, chemicals, steel, food processing	200–500 acres	N/A
3	Industrial Park on Pakistan steel mill land	Port Qasim, Sindh	Warehousing, pharmaceuticals, logistics, food processing	1500 acres	N/A
4	Rashakai special economic zone	Nowshera, KPK	Textile, manufacturing, food processing	1000 acres	Airport, railway, dry port, highway
5	Dhabeji special economic zone	Dhabeji, Sindh	Petrochemicals, automotive, FMCG, construction materials	1530 acres	Airport, railway, seaport, highway
6	Bostan special economic zone	Balochistan	Fruit processing, agriculture machinery	1000 acres	Airport, dry port, seaport, highway
7	Mirpur industrial zone	Mirpur, AJK	Multiple	1078 acres	Grand truck road, dry port, railway

(continued)

Table B (continued)

S. no.	Name	Location	Potential industries	Area	Connectivity
8	Mohmand Marble City	Mohmand, FATA	Marble industries	300 acres	N/A
9	Moqpondass special economic zone	Gilgit Baltistan	Handicrafts, mineral, livestock, Agri based	250 acres	Airport, CPEC routes, dry port

	Seven notified/approved SEZs			
S. no.	Name	Location	Potential industries	Area
1	Khairpur SEZ	Khairpur, Sindh	Agro-based industry, data processing	140 acres
2	Bin Qasim SEZ	Bin Qasim (Pak Steel Mills)	Light engineering, auto vendors, steel fabricating units, chemical and food, pharmaceuticals, electrical and consumer goods, furniture	930 acres
3	Korangi Creek	Sindh	Consumer goods, food, pharmaceuticals, garments, value-added textile, light engineering, packaging and printing, warehouses, and logistics	240 acres
4	Hattar Phase-vii	KPK	Mining, marble, fruit processing	424 acres
5	Quaid-e-Azam Apparel Park	Sheikhupura, Punjab	Textile and cotton	1536 acres
6	M-3 Industrial City	Faisalabad, Punjab	Textile, engineering and construction, chemicals, pharmaceuticals, electronics, food and beverages, information technology	4356 acres
7	Value Addition City	Faisalabad, Punjab	Textile, chemicals, pharmaceuticals, engineering, information technology	225 acres

Source: *Board of Investment (BoI) and CPEC Authority of Pakistan*

Table C Current status of pSEZs

pSEZ name	Project update
Rashakai special economic zone	• The concession agreement between the Khyber Pakhtunkhwa Economic Zone Development and Management Company (KPEZDMC) and China Road and Bridge Corporation (CRBC) was signed in April 2019. • The development agreement was signed on Sep 14, 2020, with a development cost of $128 million. • The ground breaking/launching of RSEZ was held on May 28, 2021.
Dhabeji special economic zone	• The land has been allocated. • The bidding for the development of SEZ is in process. The preferred bidder will be announced soon. • Utilities are being provided as per timelines.
Allama Iqbal Industrial City	• The ground breaking of AIIC was held on January 3, 2020, which was attended by the Prime Minister of Pakistan. The land has been acquired for AIIC, and development work has been in progress since November 2019. • Numerous national and international enterprises have been allotted plots in the SEZ. Few companies have already started construction work.
Bostan special economic zone	• The SEZ approval committee officially approved the Bostan SEZ on March 4, 2020, in the fifth Board of Approval meeting, and it was awarded SEZ status on may 5, 2020. • The development work on phase I (200 acres) has been completed. • Phase 1 of the Bostan SEZ has been launched for the allotment of industrial plots.
ICT model industrial zone	• In the pipeline.
Industrial Park on Pakistan steel mill land	• In the pipeline.
Mirpur industrial zone	• In the pipeline.
Mohmand Marble City	• In the pipeline.
Moqpondass special economic zone	• In the pipeline.

Source: *CPEC Authority of Pakistan*

References

Aggarwal, A. (2023). How special are special economic zones: Evidence from South Asia. *The World Economy, 46*(2), 2361.

Aggarwal, A., Hoppe, M., & Walkenhorst, P. (2008). *Special economic zones in South Asia: Industrial islands or vehicles for diversification?* The World Bank.

Akinci, G., & Crittle, J. (2008). *Special economic zone: Performance, lessons learned, and implication for zone development* (Vol. No. 45869, pp. 1–83). The World Bank.

Ali, M. M., & Faisal, F. (2016). CPEC, SEZ (special economic zones) and entrepreneurial development prospects in Pakistan. *The Pakistan Development Review*, 143–153.

Arbolino, R., Lantz, T. L., & Napolitano, O. (2023). Assessing the impact of special economic zones on regional growth through a comparison among EU countries. *Regional Studies, 57*(6), 1069–1083.

Arndt, C., Buch, C. M., & Schnitzer, M. E. (2010). FDI and domestic investment: An industry-level view. *The BE Journal of Economic Analysis & Policy, 10*(1).

Aslam, S., Shahid, M. N., Qureshi, M. H., & Qureshi, A. M. (2018). Investigating innovativeness and emotional intelligence as mediator to explore entrepreneurial marketing strategy focused on firm performance: A case in Pakistan. *Journal of Applied Environmental and Biological Sciences, 8*(1), 48–60.

Balasubramanyam, V. N. (1988). Export processing zones in developing countries: Theory and empirical evidence. *Economic development and international trade*, 157–165.

Blomström, M., Kokko, A., & Globerman, S. (2001). The determinants of host country spillovers from foreign direct investment: A review and synthesis of the literature. *Inward Investment Technological Change and Growth*, 34–65.

Borensztein, E., De Gregorio, J., & Lee, J. W. (1998). How does foreign direct investment affect economic growth? *Journal of International Economics, 45*(1), 115–135.

Cirera, X., Lakshman, R., & Spratt, S. (2014). The impact of export processing zones on employment, wages and labour conditions in developing countries. *3ie Systematic Review, 10.*

Dar, M. S., Ahmed, S., & Raziq, A. (2017). Small and medium-size enterprises in Pakistan: Definition and critical issues. *Pakistan Business Review, 19*(1), 46–70.

Din, M. U. (1994). Export processing zones and backward linkages. *Journal of Development Economics, 43*(2), 369–385.

Du, L., Razzaq, A., & Waqas, M. (2023). The impact of COVID-19 on small-and medium-sized enterprises (SMEs): Empirical evidence for green economic implications. *Environmental Science and Pollution Research, 30*(1), 1540–1561.

Edge, R. M., & Rudd, J. B. (2011). General-equilibrium effects of investment tax incentives. *Journal of Monetary Economics, 58*(6–8), 564–577.

Farole, T., & Akinci, G. (Eds.). (2011). *Special economic zones: Progress, emerging challenges, and future directions*. World Bank Publications.

Fingleton, B., & López-Bazo, E. (2006). Empirical growth models with spatial effects. *Papers in Regional Science, 85*(2), 177–198.

Goncharuk, A. G. (2007). Impact of political changes on industrial efficiency: A case of Ukraine. *Journal of Economic Studies, 34*(4), 324–340.

Hadi, N. U., Batool, S., & Mustafa, A. (2018). CPEC: An opportunity for a prosperous Pakistan or merely a mirage of growth and development. *The Dialogue, 13*(3).

Hamada, K. (1974). An economic analysis of the duty-free zone. *Journal of International Economics, 4*(3), 225–241.

Haque, N. U. (2007). Entrepreneurship in Pakistan, Pakistan Institute of Development Economics. PIDE Working Papers No. 29.

House, C. L., & Shapiro, M. D. (2006). Phased-in tax cuts and economic activity. *American Economic Review, 96*(5), 1835–1849.

Hussain, E., & Rao, M. F. (2020). China–Pakistan economic cooperation: The case of special economic zones (SEZs). Fudan Journal of the Humanities and Social Sciences, 13(4), 453–472.

Hyder, S., & Lussier, R. N. (2016). Why businesses succeed or fail: A study on small businesses in Pakistan. *Journal of Entrepreneurship in Emerging Economies, 8*(1), 82–100.

Iršová, Z., & Havránek, T. (2013). Determinants of horizontal spillovers from FDI: Evidence from a large meta-analysis. *World Development, 42*, 1–15.

Jiyamuratov, R. (2023). Ways to develop goods market "marketing strategies" in free economic zones. *Science and Innovation, 2*(A1), 63–69.

Johansson, H., & Nilsson, L. (1997). Export processing zones as catalysts. *World Development, 25*(12), 2115–2128.

Kankesu, J. (2003). An overview of export processing zones: Selected Asian countries. *Development Policy Review, 21*(1), 51–65.

Kemal, A. R. (2006). Key issues in industrial growth in Pakistan. *Lahore Journal of Economics, 11.*

Khan, A. A., & Khan, M. (2010). Pakistan textile industry facing new challenges. *Research Journal of International Studies, 14*(14), 21–29.

Khaskheli, M. B., Wang, S., Yan, X., & He, Y. (2023). Innovation of the social security, legal risks, sustainable management practices and employee environmental awareness in the China–Pakistan economic corridor. *Sustainability, 15*(2), 1021.

Krugman, P. (1991). Increasing returns and economic geography. *Journal of Political Economy, 99*(3), 483–499.

Lall, S. (2000). Technological change and industrialization in the Asian newly industrializing. *Technology, Learning, and Innovation: Experiences of Newly Industrializing Economies, 13.*

Lin, J. Y., & Monga, C. (2011). Growth identification and facilitation: The role of the state in the dynamics of structural change. *World Bank Policy Research Working Paper, 5313.*

Liu, Z., Zhou, J., & Li, J. (2023). How do family firms respond strategically to the digital transformation trend: Disclosing symbolic cues or making substantive changes? *Journal of Business Research, 155*, 113395.

Lu, F., Sun, W., & Wu, J. (2023). Special economic zones and human capital investment: 30 years of evidence from China. *American Economic Journal: Economic Policy, 15*(3), 35–64.

Mahmood, A., Javaid, N., Zafar, A., Riaz, R. A., Ahmed, S., & Razzaq, S. (2014). Pakistan's overall energy potential assessment, comparison of LNG, TAPI and IPI gas projects. *Renewable and Sustainable Energy Reviews, 31*, 182–193.

Parida, J. K., & Madheswaran, S. (2023). Harnessing demographic dividend before it is lost forever in India. *The Indian Journal of Labour Economics*, 1–19.

Rhee, Y. W., Katterbach, K., & White, J. (1990). *Free trade zones in export strategies* (p. 36). World Bank Industry and Energy Department. PRE, Industry Series Paper.

Rietveld, P. (1989). Infrastructure and regional development. *The Annals of Regional Science, 23*(4), 255–274.

Rodriguez, C. A. (1976). A note on the economics of the duty free zone. *Journal of International Economics, 6*(4), 385–388.

Rohra, C. I., & Panhwar, I. A. (2009). The role of SMEs towards exports in Pakistan economy. *Australian Journal of Basic and Applied Sciences, 3*(2), 1070–1082.

Sabir, I., & Sabir, A. (2010). Academic dependency of intellectual labor: The case of social sciences in Pakistan. *International Journal of Interdisciplinary Social Sciences, 5*(6).

Schrank, A. (2001). Export processing zones: Free market islands or bridges to structural transformation? *Development Policy Review, 19*(2), 223–242.

SMEDA. (2019). State of SMEs in Pakistan.

Song, Z., Zhang, G., & Dai, H. (2023). How does CVC promote disruptive innovation–case analysis based on Google and Huawei. *Asian Journal of Technology Innovation*, 1–26.

Thamwicha, W., & Chaiprasit, K. (2018). Exporting Analysis of the Small and Medium Enterprises (SMEs) in the Mukdahan Special Economic Zone. RMUTT Global Business Accounting and Finance Review, 1(2).

Warr, P. G. (1989). Export processing zones: The economics of enclave manufacturing. *The World Bank Research Observer, 4*(1), 65–88.

World Bank. (2021). World Development Indicators, the database.

World Bank FIAS report. (2008). http://sdocuments.worldbank.org/curated/en/343901468330977533/Special-economic-zone-performance-lessons-learned-and-implication-for-zone-development

Young, L., & Miyagiwa, K. F. (1987). Unemployment and the formation of duty-free zones. *Journal of Development Economics, 26*(2), 397–405.

Yuan, J. D., & Eden, L. (1992). Export processing zones in Asia: A comparative study. *Asian Survey, 32*(11), 1026–1045.

Zafar, A., & Mustafa, S. (2017). SMEs and its role in economic and socio-economic development of Pakistan. *International Journal of Academic Research in Accounting, Finance and Management Sciences, 6*(4).

Zeng, D. Z. (2016). Special economic zones: Lessons from the global experience. *PEDL Synthesis Paper Series, 1*, 1–9.

Ahsan Abbas is an Assistant Professor at the Government and Public Policy (GPP) Department, National Defence University (NDU), Islamabad, Pakistan. Prior to his current position, he worked as a Senior Research Fellow at Centre of Excellence CPEC (CoECPEC) for four years. He has also served the Planning Commission of Pakistan, the Pakistan Institute of Development Economics (PIDE), Islamabad, the National Productivity Organization (NPO), and GIFT University, Gujranwala. His professional experience as a researcher, analyst, trainer, and faculty in several public and private sector organizations ranges over 21 years. He has professional associations with prestigious national and international organizations in Economics. He holds PhD in Economics from Pakistan Institute of Development Economics (PIDE), Islamabad.

He has been associated/collaborated with various international and national organizations on various projects including Partnership for Economic Policy (PEP) Network, International Food Policy Research Institute (IFPRI), South Asia Watch on Trade, Economics & Environment (SAWTEE), South Asia Network on Economic Modeling (SANEM), United States Agency for International Development (USAID), Transparency International (TI), Sustainable Development Policy Institute (SDPI), Pakistan Institute of Trade and Development (PITAD), Pakistan Institute for Parliamentary Services (PIPS), and National Centre for Rural Development (NCRD).

He has published his research work in various recognized/reputed journals, newspapers, and books/policy documents of diverse publishing agencies. He has also presented his research outcomes in the areas of theoretical and applied economics at prestigious forums/organizations. His areas of expertise include econometrics and data analytics with application in economic development and planning, welfare economics, international trade, investment, and industrial economics. He is well versed in advanced statistical/econometric methods to analyze multifarious socioeconomic data using various mathematical/statistical/econometric software and computer packages.

Muqarab Abbas is a Ph.D. scholar in Economics at the International Islamic University, Islamabad and currently working as an Assistant Manager at the Trade Development Authority of Pakistan. Prior to his current position, he worked as an adjunct faculty member at the International Islamic University and Bahria University, Islamabad. He also worked as a researcher at the Centre of Excellence CPEC (CoECPEC), Islamabad, contributed to various publications in reputed journals and presented research work at different prestigious forums/seminars.

The Financial Well-Being of Single-Mother Entrepreneurs in the Rural Areas of Sabah, Malaysia

Salmah Topimin

Abstract The reality of women's entrepreneurship differs according to contexts. This study aims to highlight the realities of women's entrepreneurship in the context of rural areas of Sabah, Malaysia. The discussion is contextualised based on the experiences of single-mother entrepreneurs—a group rarely being highlighted within women as a heterogeneous group. The significance of entrepreneurial activities in overcoming the financial hardship faced by single mothers in rural areas is well accepted by many developing nations. This strategy is seen as relevant for single mothers to overcome the economic pressure of life while also helping to resolve the issue of poverty among single mothers. Therefore, this study explores how single-mother entrepreneurs construct their entrepreneurial experiences and the extent to which the experiences have influenced the financial well-being of their families. Semi-structured, in-depth interviews with eight single-mother entrepreneurs in the rural areas of Sabah were conducted. The idea of single mothers as a significant group in necessities entrepreneurship is established. Considering that the issue of poverty in Sabah is critical, the involvement of single mothers as necessity entrepreneurs could have an impact on the state's economic development. The financial hardship that women experience when they first become single mothers persists throughout their single motherhood lives, indicating the financial vulnerability of single-mother entrepreneurs. It is suggested that the key to achieving the financial well-being of single-mother entrepreneurs lies in a comprehensive and unique institutional strategy of its own.

Keywords Financial well-being · Poverty · Sabah · Single mothers · Women's entrepreneurship

---

S. Topimin (✉)
Faculty of Business, Economics and Accountancy, Universiti Malaysia Sabah, Kota Kinabalu, Sabah, Malaysia
e-mail: salmah@ums.edu.my

N. Hayat, N. C. Nawi (eds.), *Contemporary Issues in Entrepreneurship and Innovative Technology*, Contributions to Environmental Sciences & Innovative Business Technology, https://doi.org/10.1007/978-3-031-50927-8_2

1 Introduction

Women are one of the fastest-rising populations of entrepreneurs (Cardella et al., 2020). Regrettably, women entrepreneurs tend to be invisible in the literature due to the bias surrounding our understanding of women entrepreneurs. The bias of seeing women entrepreneurs as a homogeneous group has limited our understanding of women's entrepreneurship (Madsen et al., 2008). It should be acknowledged that women are a widely heterogeneous group with different needs and challenges that shape their entrepreneurial journey. It is believed that by recognising women as a heterogenous group, the bias against women entrepreneurs can be avoided in at least two situations. First, women entrepreneurs would not be viewed from the experience of men (Ahl, 2006; Loscocco & Bird, 2012). Second, women entrepreneurs would be viewed based on the understanding that there are different kinds of women entrepreneurs with different needs and aspirations and who are affected differently by their social environments (Brush & Gatewood, 2008).

This study addresses this bias by investigating the entrepreneurial activities of single mothers and the extent to which they have achieved their family's financial well-being. The investigation is contextualised based on the experience of single-mother entrepreneurs in the rural areas of Sabah, Malaysia. Therefore, this study acknowledges single mothers as a significant group of women entrepreneurs. In this respect, this study has responded to the need for conducting research on women that recognises the differences among groups of women (Brush & Greene, 2021). In addition, the present study responds to the call for research on the economic activities and financial vulnerability of single mothers, thus incorporating their discussion into the mainstream of entrepreneurship and public debates (Matsai & Raniga, 2021). Therefore, this study explores the following research question:

> How do single-mother entrepreneurs in the rural areas in Sabah perceive and view their entrepreneurial activities, and to what extent do the activities influence their financial well-being?

With these objectives in mind, the chapter is structured as follows. The subsequent section provides an overview of Malaysian women's participation in the labour force, followed by an overview of single mothers in Sabah as the research context. Next, the reviewed literature on women's entrepreneurship is presented. The subsequent section discusses the methodology used for data collection. The chapter concludes with a discussion of the study's findings and recommendations to strengthen the research on women's entrepreneurship and improve the strategies targeted towards single mothers in general and single-mother entrepreneurs in particular. The findings may contribute to the existing literature on women and women's entrepreneurship by highlighting the reality of the financial well-being of single-mother entrepreneurs in rural areas. Therefore, a better entrepreneurial ecosystem for single mothers can be proposed.

2 Context

In 2022, the women's population in Malaysia was recorded at 15.7 million, representing 47.7% of the total population (Department of Statistics Malaysia, 2022), indicating the significance that women can have in the country's economic development. One aspect of economic development that is relevant for women is through their involvement in the labour force. However, the gender gap in the labour force participation rate has been a long-standing issue in Malaysia. As demonstrated in Table 1, the labour force participation rate of men and women in Malaysia differs significantly. For the past 5 years, women's participation in the labour force is still far behind men's.

Regrettably, it has historically been reported that the labour participation rate is lower for Malaysian women than for men. Nonetheless, an impressive improvement shows that more women have been participating in the labour market. For example, it was reported that women's labour force participation rate between 2020 and 2017 increased by 8%, nine times more than the 0.9% increase recorded for the period between 1995 and 2010 (Khazanah Research Institute, 2018). Since the ratio of Malaysian women's involvement in tertiary education has steadily increased over the years (Asadullah, 2019), the increased number of women in the Malaysian labour force is not surprising. However, women's involvement in the Malaysian labour force is likely to place women at a disadvantage in at least two situations. First, women are mostly employed in feminine-related jobs. For example, most women are employed as service and sales workers, clerical support workers, and professionals. However, although the percentage of women as professionals is higher than men, women are more likely to work in the healthcare industry as nurses, midwives, or teaching professionals. By contrast, for other professional jobs, such as legislators, senior officials, and managers, women (24%) were still lagging behind men (76%) (Department of Statistics Malaysia, 2019). This situation indicates that the issue of gender segregation in the occupational structure among Malaysian women persists, placing women as secondary to men. Second, the employment status of women in Malaysia, in contrast to that of men, also demonstrates that they are more frequently employed as own-account workers and unpaid family workers than as employees and employers. Generally, the employment status in Malaysia is categorised into four different types: employer, employee, own-account worker, and unpaid family helper. Within this context, the percentage of own-account workers, also known as self-employed

Table 1 Labour force participation rate of women in Malaysia between 2017 and 2021

	Labour force participation rate (%)				
	2021	2020	2019	2018	2017
Female	55.5	55.3	55.8	55.3	53.5
Male	80.9	80.6	80.8	80.5	77.7

Source: Compile from a wide range of official Department of Statistics, Malaysia, reports

workers, has increased drastically over the years. A report by Khazanah Research Institute (2018) that analysed the changes in women's labour participation in Malaysia between 2010 and 2017 stated that the increased number of women's labour force for the two periods was contributed by the increased number of women as own-account workers. In fact, among the four types of employment status in Malaysia, it is the women own-account workers who grew the most. This situation may indicate that women became more active in participating in the country's economic activities. However, there is also a concern that those who become own-account workers actually struggle to find a place in the labour market, thus forcing themselves to become own-account workers with unstable income and limited business opportunities (Schaper, 2020). At this point, there is a possibility that the issue of women's economic vulnerability still remains even if they are actively participating in economic activities.

One significant economic activity of women as own-account workers is entrepreneurial activities. Various factors can be associated with the growth of women-owned establishments in Malaysia. Interestingly, while contextual and institutional aspects (e.g. culture, gender roles) are discussed as limiting women's potential in their productive activities, the same aspects are also crucial in influencing women towards entrepreneurial activities (e.g. flexibility). Topimin (2021) suggests that women may become self-employed because they have no practical career options, particularly for women who are the primary income earners of their families. One category of women who become the primary earners of households is single mothers. Being a single mother is normally associated with a misfortunate situation that made a woman the provider and nurturer of her family.

Consequently, it becomes a priority for single mothers to balance their domestic and work responsibilities (Kariv, 2013). Since single mothers are discussed as lacking work-related skills to compete for employment opportunities (Blank, 2007), creating their own careers through entrepreneurial activities becomes necessary. However, it is never easy for single-mother entrepreneurs to maximise their entrepreneurial potential. Often, their business performance is highlighted as lower than the achievement of other women entrepreneurs (e.g. single or married women) (Salwa et al., 2015). Therefore, it is the interest of this chapter to investigate the extent to which single-mother entrepreneurs who are categorised as own-account workers have achieved their family's financial well-being.

3 Single Mothers in Sabah

There is an increase in the number of single mothers worldwide (Adellyan Rasi et al., 2013), including in Malaysia (Ahmad et al., 2017). A similar trend is also observed in Sabah with regard to the increased number of single mothers. Data from the Malaysia Population Census in 2000 and 2010 show that the number of single mothers in Sabah increased from 46,859 in 2000 to 61,717 in 2010 (Department of Statistics Malaysia, 2000, 2010), thus reflecting the increased number of female-headed households in the state. Within this context, female-headed households in Sabah have several distinctive characteristics, as shown in Table 2.

Table 2 shows that female-headed households in Sabah were represented mostly by widows and divorcees and women who have separated from their husbands, indicating their positions as single mothers. In addition, although single mothers can be found in most of the age groups, it is worth noting that they also fall under the working age category (15–64), suggesting that they are significant contributors to the state's labour force and may experience stress issues while performing their productive and domestic roles. Regarding the employment status of female-headed households in Sabah, data from the Malaysian Census 2010 (Department of Statistics Malaysia, 2010) show that women make up a smaller percentage of employers and employees in both the public and private sectors than males. In addition, among occupation categories highlighted in the census, a significant percentage of female-headed households are self-employed (13.6%). Within this context, getting involved in entrepreneurial activities has been discussed as a common way for single mothers to generate income (Abdullah et al., 2015). However, it is argued that single mothers in Sabah encounter issues such as having irregular income and high levels of job insecurity (Mulia, 2017; Topimin et al., 2019), indicating their financial vulnerability, given that Sabah is one of the three Malaysian states with the highest rates of poverty (Habibullah et al., 2018) in both rural and urban areas (Department of Statistics Malaysia, 2016). It is more likely that this circumstance will also affect the financial well-being of single-mother entrepreneurs in the state.

Table 2 Female-headed households by age and marital status in Sabah

	Marital status				
Age group	Never married (%)	Currently married (%)	Widowed (%)	Divorced (%)	Separated (%)
20–24	34.3	16.1	–	–	–
25–29	30.1	8.9	62.5	62.5	60.0
30–34	23.7	6.2	55.6	66.7	60.0
35–39	35.1	6.2	67.7	82.6	44.4
40–44	32.5	6.6	73.6	92.0	22.2
45–49	37.8	6.1	77.0	85.7	45.5
50–54	41.5	6.7	76.5	50.0	60.0
55–59	47.1	5.9	74.2	64.7	25.0
60–64	64.3	10.2	74.3	66.7	100.0
65–69	–	12.5	72.3	33.3	100.0
70–74	25.0	8.5	75.4	33.3	50.0
75 +	25.0	7.3	56.8	–	100.0

Source: Computed with 2% sample data from the population census of Malaysia 2010

4 Single Mothers and Entrepreneurship Activities

The need to not view women as a homogenous group has been widely acknowledged in women's literature (Ahl, 2006; Moser, 1993). The tendency to view women as a homogenous group may discriminate women against men as various factors influencing women's behaviour and interactions in both public and private spheres could not be revealed. It should be acknowledged that women's needs vary widely and are determined by various factors, such as their socioeconomic conditions (e.g. educational level and occupation) and socio-cultural aspects (e.g. social status and social values). In addition, the extent to which women can successfully interact and act in the public sphere also depends on how the social environment categorises them. For example, a woman's marital status (a socio-demographic variable) may impact how much she can participate in productive activities. In fact, it has long been argued that the economic conditions of women who are the heads of their households vary considerably according to factors such as a woman's marital status and her access to productive resources and income (Moser, 1993). Within the heterogeneity of women as a group, the realities experienced by single-mother entrepreneurs are rarely explored. More often than not, the financial vulnerability of single mothers is prevalent, yet there is a lack of research to reveal the phenomenon.

Single mothers face various challenges in entering the labour market. They are subject to various limitations (e.g. skills, educational attainment, networking, multiple gender roles) that make it possible for them to compete fairly for job opportunities. As a result, one approach to ensure that they can support their family's needs is to engage in entrepreneurial activities (Abdullah et al., 2015), particularly by operating small businesses that rely on women's specific skills, such as sewing, crafting, and cooking. Such actions are called necessity entrepreneurship—when an individual enters self-employment by creating a small venture due to no other employment option (Nikiforou et al., 2019). It is not uncommon that motivational factors behind women's entrepreneurship are rooted in many aspects of women's frustration and dissatisfaction in performing their roles in public and private spheres. As such, it is unsurprising that women are more likely than men to become necessity entrepreneurs in most nations (Kelley et al., 2010). This argument applies practically to single-mother entrepreneurs.

Life as single mothers is never easy as they are equally responsible for both work and domestic duties. Like other women entrepreneurs (e.g. married and single women), single mothers find it challenging to run their own enterprises. These challenges stem from various aspects of women's lives. First, as widely discussed in the women's entrepreneurship literature, women entrepreneurs are constrained with financial resources (Alsos et al., 2006). The fact that women are more likely to have had prior employment that provides them with lower income, Rusyda et al. (2011) has indirectly reduced the amount of capital that women can contribute to starting a business. For single-mother entrepreneurs, the under-capitalisation issue becomes more significant as not many have been involved in productive activities before becoming single mothers. Second, women entrepreneurs also lack business-management-related skills that are significant in facilitating their business operations (Blank, 2007). In general, single mothers gain experience and knowledge through their interactions and connections within informal social networks, specifically with family, friends, and the local community. Even if a single mother has formal interaction through work experience, that experience still adds a little to their business activities because women tend to be involved in women-related jobs. Therefore, women's employment-related experience also provides them with inadequate knowledge that can be used to realise the potential of their entrepreneurial endeavours fully. Third, the need for single mothers to carry out dual responsibilities as the providers and nurturers of their families has also placed women in a difficult situation. The problem of balancing multiple roles is exacerbated for single mothers, affecting the outcome of their business activities. While the involvement of single mothers in creating business activities is hoped to empower their economic status, previous research shows that their business is underperformed and is easily exposed to the risk of business failure (Abu Bakar et al., 2004). As such, it is highlighted that financial issues become a source of major stress for single-mother entrepreneurs (Hashim et al., 2015).

Financial well-being refers to people's satisfaction with their financial situation. Financial satisfaction is conceptualised as possessing a healthy financial position and is strongly related to the level of income, savings, and other material security that someone has (Sabri & Falahati, 2012). Theoretically, financial well-being is one of the elements used in measuring the achievement of the well-being of a nation. In the Malaysian context, financial well-being has been included in measuring the country's well-being index in two ways: first, the use of financial indicators such as income to measure the economic well-being of its nation and, second, financial well-being indicators that are used to measure the family economy of Malaysian households. The latter is the concept of financial well-being used for the discussion in this study. Based on the National Population and Family Development Board of Malaysia (NPFDB), the family economy covers financial well-being and financial management (NPFDB, 2016). The first domain of financial well-being refers to the ability of a family to meet the current financial well-being and settle existing financial commitments. In contrast, the second domain refers to how well a family manages its financial resources in accordance with the financial goals of the family. Although a family's well-being in Malaysia is measured through various domains, perhaps it is not too much to conclude that financial well-being has a big impact on the life of single mothers in Sabah, particularly when they are closely linked with the issue of poverty. For example, it is highlighted that the incidence of poverty among single mothers in the rural areas of Sabah is profound (Topimin, 2021). It was also reported that a significant number of single mothers in the rural areas of Sabah were living below the poverty line (Mulia, 2017). In this respect, there is a possibility that single mothers in Sabah could be financially vulnerable due to their distinctive location in rural areas, suggesting a need for further investigation on this aspect.

5 Methodology

To generate a holistic view of the financial well-being of single-mother entrepreneurs in Sabah, this study adopts a qualitative research design that capitalises on semi-structured, in-depth interviews (Creswell, 2014). The qualitative approach emphasises an in-depth understanding and the details of the phenomenon being examined and requires involvement from researchers at all stages of the research process (Yoosefi Lebni et al., 2020). The need to apply constructivist inquiry in researching women is significant to limit the views of the researcher and to provide excellent opportunities for single-mother entrepreneurs to share their opinions about their entrepreneurial activities and the extent to which the activities affect their financial well-being without being controlled by the researcher's technical procedure (Bryman, 2016). In addition, the issue of gendered entrepreneurship (Cardella et al., 2020) can be minimised as the experience of single-mother entrepreneurs is investigated based on their perceptions, views, knowledge, and interpretations. This practice is consistent with the long-standing argument that priority should be placed on replacing the researchers' views with the women's views when researching women (Mies, 1993).

An interview guide was used to ensure a systematic process while conducting interviews with different single-mother entrepreneurs. However, since it is almost impossible to determine how participants will answer the interview questions, the researcher was free to probe and ask questions spontaneously (Patton, 2015). Data for this study were collected based on in-depth interviews conducted with eight single-mother entrepreneurs in two rural areas in Sabah, Malaysia. This study is at a preliminary stage, with a focus on investigating the status of the financial well-being of single-mother entrepreneurs in the selected areas before a thorough investigation is carried out throughout the entire state of Sabah. As such, the study only analysed data from a small number of participants. However, the selection of eight single mothers in this study is in accordance with the inclusion criteria established by the researcher, in which they must be single mothers who are actively running a business and the business serves as their primary source of income. These criteria help the researcher ensure that the single mothers who participated in this study are fully and actively involved in their entrepreneurial activities, thus providing information-rich cases (Patton, 2015). Although this study uses a small sample size, data from the sample reached the theoretical saturation point (Straus & Corbin, 1998). One possible explanation for obtaining repetitive data from the small sample size in this study could be due to the similarity of the entrepreneurial activities conducted and the geographical context of the participants, providing them with similar entrepreneurial experiences.

In-depth interviews, which lasted between 60 and 90 min, with eight single-mother entrepreneurs were conducted and recorded with their consent. The interviews focused on three key areas: the demographic profiles of single-mother entrepreneurs, the motivations for business, and the status of single-mother entrepreneurs' financial well-being. The recorded interviews were then transcribed and analysed under Braun and Clarke's (2006) thematic analysis. The data analysis starts with some priori themes or topics identified beforehand. For example, some issues that define the financial well-being of single mothers have been studied in the literature, such as low income. Thus, it is anticipated that this theme will emerge in the data. However, the pre-identified codes or themes may be changed throughout the research to address better the problems that the data raised. Based on their differences and similarities, codes and themes were categorised.

6 The Motivations for Business

The categories of single mothers who were involved in this study are widows (5) and divorcees (3). The findings show that divorced single mothers are often younger (under 50) than widowed single mothers (over 50). Three categories of age were used to characterise single mothers in this study (30–39, 40–49, and above 40). Based on these categories, the data show that older single mothers have been the head of the family for longer than the younger age group. Their socio-demographic profile shows that only one single-mother entrepreneur obtained a university education, whereas the other seven completed secondary schooling. The entrepreneurial activities of single mothers can be found in two main sectors: service and agriculture. Seven of the single-mother entrepreneurs have established their own ventures after they became single mothers, in which the business is developed based on their home-related skills. Several of them are involved in small-scale farming activities, where they utilise their family's land (e.g. parents). All single-mother entrepreneurs in this study became single mothers by force and not by choice, significantly impacting their incapacity to cope with their lives. The analysis reveals that push factors drive all single mothers into their business ventures. The main factor influencing the decision of single mothers to engage in entrepreneurial activities is to ensure their survival.

For example, SM3, who was married at 17 and became a single mother at the age of 29 with six children, stated:

> I must find ways to survive, provide for my children's necessities, and secure their future. Back then, things were quite challenging. I used to work as a part-time cleaner, but the payment received was inadequate to support my family. Luckily, I learned massage skills from the elders, and people have given me positive feedback on this talent. I have good feedback from others about this talent. So … I focus on this for my income ever since.

Another single mother who was married at the age of 17 and became a single mother 4 years ago with two children stated:

> I used to work in a restaurant but stopped after having my first child. Things were not easy when I got divorced from my husband because I had only finished secondary school and had two children. I had to go back to my parent's house. There is nowhere else to go but to my parent's home. However, they are poor as well. In this village, there are many farmers, so with the little money I got after the divorce, I bought vegetables from the locals and sold them here next to the main road. That is how I provide for my family.

Divorcees are more likely to struggle than widows to provide for their families, especially since they do not have a house after the divorce. For widowed single-mother entrepreneurs, they do have a place to reside, but because they tend to be older, it is difficult for them to find jobs to support their lives. SM7, who became a widow in the mid-40s and had four children, ranging in age from secondary school age to early tertiary education age, stated:

> My husband passed away when all my four kids were still in their education. I had never been working while my husband was around. My life is much better than some of my friends. I got a house and savings left by my late husband. But I have to work … I do not want my children's education affected after losing their father. I know that my savings will eventually run out … so I must start generating income. Of course, looking for a job that fits my age and situation at that time is not easy. I have no previous working experience, and finding a job in my mid-forties without specific skills … it does not make sense.

7 The Status of Single Mothers' Financial Well-Being

This study reveals that all participating single-mother entrepreneurs have mixed feelings about their financial well-being. On the one hand, single-mother entrepreneurs expressed their dissatisfaction and worried about their financial situation. Being regarded as necessity entrepreneurs indicates that single-mother entrepreneurs have a limitation in maximising the potential of their entrepreneurship activities. This situation can be seen from the terms single mothers use when describing the financial return of their businesses. The following interview extracts reflect on their negative feelings about their income:

> … as long as I can provide food for my family (SM1)

> There is a time when I have to borrow money from my friends, especially for my working capital (SM3)

> I have been a tailor since my husband passed away. However, the income is not so high because of the limited demand from people in this village. So, to add income … I sell local cakes as well (SM4)

> I do many things … tailoring … farming. Why? To make sure that I can fulfil the needs of my family (SM5)

> I grow chilli on a small plot of land behind the home … of course, the income is not much (SM6)

> I used to grow chilli … but now I am more on selling foods. The income is enough for my survival (SM7)

> As for now, okay … but to say that I am financially stable … I am not sure about that (SM8)

Except for SM2, who inherited the business from her late husband, the other single-mother entrepreneurs are dissatisfied with the income they generate from their activities.

On the other hand, it is apparent that at the beginning of their single motherhood stage, most of the single mothers in this study had to struggle to fulfil their families' basic needs, such as providing food, shelter, and education for their children. When single-mother entrepreneurs share their views on this matter, they concur that what they have today is way better than what they had experienced before. The financial hardship has caused single mothers to feel grateful for what they have received and created a sense of "better than nothing". For example, single-mother entrepreneurs stated:

> I am constantly appreciative of what I have now. Maybe my money is nothing to some folks. However, the fact that I have food on the table, my children can finish their schooling, and I am self-sufficient is what matters most to me. Everything was challenging when I first became a single mother, so I should be grateful that I have made it this far (SM7).

> Everyone desires more, am I right? A lot of wealth and money, but what can I do? I am thankful for what I do have. Not much ... but I can still survive and go on with my life (SM1).

Meanwhile, there are times when single-mother entrepreneurs must turn to their relatives and close friends for financial assistance, particularly when it comes to meeting some of their financial obligations. The most common situation that has forced them to borrow money from family and friends is when they are dealing with the financial requirements for their dependents' health and education. Pertaining to the management of financial resources by single-mother entrepreneurs, they all expressed that they have plans on how their income should be distributed. They agree that their income is allocated to *foods*, *utilities*, *education costs*, and *other household expenses*. Regrettably, although *savings* is also part of their financial management, it is difficult for single-mother entrepreneurs to maintain a secured level of savings. Other types of security, like owning fixed assets, are irrelevant.

8 Discussion and Conclusion

The findings revealed that single mothers have similar entrepreneurial experiences that affect their financial well-being and livelihood. Becoming single mothers has forced the women in this study to initiate their own small business activities, as highlighted by Abdullah et al., 2015. The views shared by single-mother entrepreneurs in this study show that they were solely being motivated by push factors of the need to generate income to support the livelihood of their families. In this sense, they are being driven into entrepreneurship by the influence of a negative environment, particularly in placing themselves in the labour market. This situation makes their entrepreneurial activities fit nicely under necessity entrepreneurship (Nikiforou et al., 2019). In general, the discussion of necessity entrepreneurship always relates to unemployed people and becomes particularly relevant among single mothers. However, the complexity of roles that single mothers have to perform as the head of their households and the caregivers of their families (Sheikh et al., 2018) provides a different reality of necessity entrepreneurship from that of other unemployed people (e.g. unemployed men). With a conducive entrepreneurial landscape provided by policymakers, switching roles between necessity and opportunity entrepreneurs can be made possible. For example, unemployed people might begin their entrepreneurial activities as necessity entrepreneurs but later become opportunists and switch roles to opportunity entrepreneurs (Boyce et al., 2015). Regrettably, the findings of this study highlight that in addition to the limited amount of human capital (e.g. low level of education and a lack of working experience prior to their single motherhood) and financial capital possessed by single mothers, they struggle with some basic living necessities, such as shelter and food. More often than not, single-mother entrepreneurs would remain as necessity entrepreneurs due to the various business and non-business limitations and challenges they face. As such, the financial hardship women experience when they first become single mothers persists throughout their lives. In this sense, this study provides another piece of empirical evidence on the issue of poverty, financial hardships, and vulnerability among single mothers (Mulia, 2017; Topimin et al., 2019).

In addition, this study provides evidence that single mothers rely on informal support for their business activities. Although relying on informal support is a long-standing issue for women entrepreneurs in general (e.g. Roomi et al., 2009), this issue is even more acute for single-mother entrepreneurs. Undeniably, informal support is useful; however, this study found that this support cannot help single mothers achieve their financial well-being and reduce their financial hardships. Within this context, this study revealed that the financial support received from family and friends is meant to meet non-business needs (e.g. the health and education of dependents). The most common business-related resource that single mothers receive in this study is land that their families granted them permission to use for farming activities. However, inadequate financial support has limited the income generated from their farming activities. In this sense, this finding confirms that the key to the financial well-being of women in rural areas involved in income-generating activities through farming is the availability of formal financial services and support that meets their needs (OECD, 2021).

In essence, it is evident that self-employment through entrepreneurial activities is the prevailing employment option for single mothers in the rural areas of Sabah. Since single mothers lack job-related skills and experience, the choice of being self-employed is relevant. However, while women's self-employment aims to help women achieve empowerment, this is not true for single-mother entrepreneurs. Regrettably, the financial hardship women experience when they first become single mothers persists throughout their lives. As a consequence, this study highlights the financial vulnerability of single-mother entrepreneurs. The notion of sustaining financial well-being is far from what they have accomplished monetarily. Finally, the issue of the financial well-being of single-mother entrepreneurs not only affects the well-being of their families but most importantly contributes to the development of a well-functioning society of a nation. As such, ensuring the well-being of women is important for both individual women and various levels of institutions (i.e. family, household, society, and nation).

9 Recommendations

Having discussed the entrepreneurial experiences of single mothers in rural areas of Sabah, this study offers recommendations to improve the outcome of their entrepreneurial activities and reduce their financial vulnerability. Specifically, this study recommends that a comprehensive and unique institutional strategy is the key to achieving single-mother entrepreneurs' financial well-being. In this respect, any effort to support the initiation and survival of single-mother entrepreneurs' businesses should be provided after a thorough analysis of their specific needs. Evidently, the single mothers in this study are necessity entrepreneurs motivated to generate income to support their livelihood. In this respect, unlike other groups of women who may be positively driven to entrepreneurial careers, it is otherwise for single mothers. Therefore, the main priority of any support programme for single mothers must be to foster their entrepreneurial spirit and mindset. With this focus, it is hoped that such programmes can unleash their entrepreneurial potential and enable them to move up from being necessity to opportunity entrepreneurs, which would help strengthen their financial well-being. In addition, considerable attention should be given to designing support programmes for single mothers.

First and foremost, any programmes designed for single mothers should not be provided based on the concept of the welfare approach. The programmes should be able to facilitate single mothers to build up their business skills and capabilities and not treat them as recipients of social benefits. In this sense, the support should be provided as a means to an end, a way to get a better livelihood for single mothers. Finally, this study has shown that single mothers in rural areas are involved in entrepreneurial activities that utilise their home-related skills, a common practice for women in general. Considering the importance of farming activities in rural areas in the Sabah context, it is suggested that support providers should offer comprehensive entrepreneurial support programmes that reflect on various challenges and barriers faced by single mothers in farming activities.

Most importantly, the support system designed for single-mother entrepreneurs in rural areas must be based on the adaptive approach and consider gender sensitivity issues affecting women in general and single mothers in particular. Applying the adaptive approach in the support programmes for single mothers should be incorporated into the design and implementation of such programmes, thus ensuring their effectiveness. If a one-size-fits-all approach does not work effectively for women's support programmes, applying it to single mothers will make the issue of the ineffectiveness of such programmes even worse.

10 Limitations and Suggestions for Future Research

This study is not without limitations. It uses a qualitative approach to explore the views of single-mother entrepreneurs in relation to their entrepreneurial experiences and financial well-being. This approach is considered appropriate for exploring topics with vulnerable groups such as single mothers, though it may be useful for future research to consider a mixed-method approach. While qualitative data help increase the understanding of the reality of single-mother entrepreneurs through their own voices, statistical findings, on the other hand, would be useful to support recommendations for policymakers and provide statistical figures for future research. This study was conducted based on the small sample size of single-mother entrepreneurs in two selected rural areas in Sabah. Further research that captures other rural areas in Sabah with a larger sample would be beneficial in providing a better picture of the financial well-being of single-mother entrepreneurs, thus enhancing the validity of research findings.

Acknowledgements The Fundamental Research Grant Scheme, Ministry of Higher Education, Malaysia (FRGS/1/2020/SS01/UMS/02/1), supported this research.

References

Abdullah, J., Yeop Johari, H. S., Ahmad Bakri, A., & Wan Abdul Razak, W. M. (2015). Students and women entrepreneurs' collaborations in social enterprise program at UITM, Malaysia. *Procedia Social and Behavioral Sciences, 168*, 97–103. https://doi.org/10.1016/j.sbspro.2014.10.214

Abu Bakar, T. S., Omar, R., Md. Ali, A., Md. Som, H., Mukhtar, S. N., & Mohd Jamil, R. (2004). *Entrepreneurial activities among single mothers in Johor*. Universiti Teknologi Malaysia.

Adellyan Rasi, H., Moula, A., Puddephatt, A. J., & Timpka, T. (2013). Empowering single mothers in Iran: Applying a problem-solving model in learning groups to develop participants' capacity to improve their lives. *British Journal of Social Work, 43*(5), 833–852. https://doi.org/10.1093/bjsw/bcs009

Ahl, H. (2006). Why research on women entrepreneurs needs new directions. *Entrepreneurship Theory and Practice, 30*(5), 595–621. https://doi.org/10.1111/j.1540-6520.2006.00138.x

Ahmad, S. Y., Sabri, M. F., Abd Rahim, H., & Osman, S. (2017). Factors predicting financial security of female-headed households. *Journal of Emerging Economies and Islamic Research, 5*(1), 25–38. https://doi.org/10.24191/jeeir.v5i1.8794

Alsos, G. A., Isaksen, E. J., & Ljunggren, E. (2006). New venture financing and subsequent business growth in men and women-led businesses. *Entrepreneurship Theory and Practice, 30*(5), 667–686. https://doi.org/10.1111/j.1540-6520.2006.00141.x

Asadullah, M. N. (2019). The changing status of women in Malaysian society. In N. P. Tey, S. L. Lai, & N. A. Ismail (Eds.), *Population situation analysis Malaysia 2018* (pp. 102–126). Population Studies Unit, University of Malaya.

Blank, R. M. (2007). Improving the safety net for single mothers who face serious barriers to work. *The Future of Children, 17*(2), 183–197.

Boyce, C. J., Wood, A. M., Daly, M., & Sedikides, C. (2015). Personality change following unemployment. *Journal of Applied Psychology, 100*(4), 991–1011. https://doi.org/10.1037/a0038647

Braun, V., & Clarke, V. (2006). Using thematic analysis in psychology. *Qualitative Research in Psychology, 3*(2), 77–101. https://doi.org/10.1191/1478088706qp063oa

Brush, C. G., & Gatewood, E. J. (2008). Women growing businesses: Clearing the hurdles. *Business Horizons, 51*(3), 175–179. https://doi.org/10.1016/j.bushor.2008.01.007

Brush, C. G., & Greene, P. G. (2021). Do women engage differently in entrepreneurship? In T. M. Cooney (Ed.), *The Palgrave handbook of minority entrepreneurship* (1st ed., pp. 139–158). Springer.

Bryman, A. (2016). *Social research methods* (5th ed.). Oxford University Press.

Cardella, G. M., Hernández-Sánchez, B. R., & Sánchez-Garcia, J. C. (2020). Women entrepreneurship: A systematic review to outline the boundaries of scientific literature. *Frontiers in Psychology, 11*, 1–18. https://doi.org/10.3389/fpsyg.2020.01557
Creswell, J. W. (2014). *Research design: Qualitative, quantitative and mixed methods approaches* (4th ed.). Sage Publications.
Department of Statistics Malaysia (DOSM). (2000). Census 2000.
Department of Statistics Malaysia (DOSM). (2010). Census 2010.
Department of Statistics Malaysia (DOSM). (2016). Household income survey 2009–2016.
Department of Statistics Malaysia (DOSM). (2019). Statistics on women empowerment in selected domains, Malaysia, 2022.
Department of Statistics Malaysia (DOSM). (2022). Current population estimates, Malaysia, 2022.
Habibullah, M. S., Sanusi, N. A., Abdullah, L., Kusairi, S., Hassan, A. A. G., & Ghazali, N. A. (2018). How long does it take for a poor state to catch up to a richer state in Malaysia? A note. *International Journal of Business and Society, 19*(2), 269–280.
Hashim, I. H. M., Azmawati, A. A., & Endut, N. (2015). Stress, roles and responsibilities of single mothers in Malaysia. *SHS Web of Conferences, 18*, 1–8. https://doi.org/10.1051/shsconf/20151803003
Kariv, D. (2013). *Female entrepreneurship and the new venture creation: An international overview*. Routledge Taylor & Francis Group.
Kelley, D. J., Brush, C. G., Greene, P. G., & Litovsky, Y. (2010). *Global entrepreneurship monitor 2010 women's report*. Babson College.
Khazanah Research Institute. (2018). The state of households 2018 – Different realities. Retrieved January 3, 2023, from https://www.krinstitute.org/assets/contentMS/img/template/editor/FullReport_KRI_SOH_2018.pdf
Loscocco, K., & Bird, S. R. (2012). Gendered path: Why women lag behind men in small business success. *Work and Occupations, 39*(2), 183–219. https://doi.org/10.1177/0730888412444282
Madsen, M. T., Neergaard, H., & Ulhøi, J. P. (2008). The influence of roles and identities on female entrepreneurial agency. *International Journal of Entrepreneurship and Small Business, 5*(3–4), 358–372. https://doi.org/10.1504/IJESB.2008.017309
Matsai, V., & Raniga, T. (2021). Economic stressors and coping strategies of single mothers living in poverty in Zimbabwe. *Social Work, 57*(2), 182–192.
Mies, M. (1993). Towards a methodology for feminist research. In M. Hammersley (Ed.), *Social research: Philosophy, politics and practice* (pp. 64–82). Sage Publications.
Moser, C. O. N. (1993). *Gender planning and development: Theory practice and training*. Routledge.
Mulia, D. S. (2017). Survival strategies of single mothers among indigenous ethnics in rural areas: Case study in Kota Belud, Sabah. *Jurnal Kinabalu, 23*, 43–63.
National Population and Family Development Board of Malaysia NPFDB. (2016). Report on key findings of the fifth Malaysian population and family survey (MPFS-5) 2014.
Nikiforou, A., Dencker, J. C., & Gruber, M. (2019). Necessity entrepreneurship and industry choice in new firm creation. *Strategic Management Journal, 40*(13), 2165–2190. https://doi.org/10.1002/smj.3075
OECD. (2021). Strengthening women's entrepreneurship in ASEAN countries. Retrieved March 31, 2023, from https://www.oecd.org/southeast-asia/regional-programme/networks/OECD-strengthening-women-entrepreneurship-in-agriculture-in-asean-countries.pdf.
Patton, M. Q. (2015). *Qualitative research and evaluation methods: Integrating theory and practice* (4th ed.). Sage.
Roomi, M. A., Harrison, P., & Beaumont-Kerridge, J. (2009). Women-owned small and medium enterprises in England: Analysis of factors influencing the growth process. *Journal of Small Business and Enterprise Development, 16*(2), 270–288. https://doi.org/10.1108/14626000910956056
Rusyda, H. M., Lukman, Z. M., Subhi, N., Chong, S. T., Abdul Latiff, A., Hasrul, H., & Wan Azimah, W. M. (2011). Coping with difficulties: Social inequality and stigmatisation on single

mothers with low-income household. *Pertanika Journal of Social Sciences & Humanities, 19*, 157–162.

Sabri, M. F., & Falahati, L. (2012). Estimating a model of subjective financial well-being among college students. *International Journal of Humanities and Social Science, 2*(18), 191–199.

Salwa, A. H. F., Yahya, M. H. M., Azila, A. R., & Fidlizan, M. (2015). The performance of married and single-mother entrepreneurs: Evidence from Malaysia. In R. Omar, H. Bahrom, & G. de Mello (Eds.), *Islamic perspectives relating to business, arts, culture and communication* (pp. 419–428). Springer.

Schaper, M. T. (2020). Malaysia's self-employment explosion: Why so many own-account workers? Retrieved January 7, 2023, from https://www.iseas.edu.sg/wp-content/uploads/2020/02/ISEAS_Perspective_2020_20.pdf

Sheikh, S., Sist, F., Akdeniz, A., & Yousafzai, S. (2018). Developing an understanding of entrepreneurship intertwined with motherhood: A career narrative of British mumpreneurs. In S. Yousafzai, A. Lindgreen, S. Saeed, C. Henry, & A. Fayolle (Eds.), *Contextual embeddedness of women's entrepreneurship: Going beyond a gender-neutral approach* (pp. 219–232). Routledge.

Straus, A., & Corbin, J. (1998). *Basic of qualitative research: Techniques and procedures for developing grounded theory* (2nd ed.). Sage Publications.

Topimin, S. (2021). Women, family, and gender issues. In N. P. Tey, S. L. Lai, & J. K. L. Chan (Eds.), *Demographics and socioeconomics changes in Sabah* (pp. 259–281). Penerbit UMS.

Topimin, S., Fabeil, N. F., & Abdullah, A. S. (2019). Women's business survival: Challenges and strategies for single mother entrepreneurs. *Academic Journal of Business and Social Sciences, 3*(Special Edition), 1–10.

Yoosefi Lebni, J., Mohammadi Gharehghani, M. A., Soofizad, G., Khosravi, B., Ziapour, A., & Irandoost, S. F. (2020). Challenges and opportunities confronting female-headed households in Iran: A qualitative study. *BMC Women's Health, 20*, 183. https://doi.org/10.1186/s12905-020-01046-x

Salmah Topimin is a senior lecturer at the Faculty of Business, Economics, and Accountancy, Universiti Malaysia Sabah. She obtained her Ph.D. in entrepreneurship from Nottingham Trent University, UK 2015. Her research area particularly focuses on women's entrepreneurship.

Enviropreneurial Orientation and Industry 4.0: Mediating the Role of Enviropreneurial Marketing in an Emerging Economy

Kashif Ali, Azizan Osman, Thavamaran Kanesan, Liang Rui, Reem Abdulrahman Alfaisaly, and Abdulrahman Saleh Alalyani

Abstract Customers are willing to pay extra for environmentally friendly goods and services, which is providing new challenges and opportunities for entrepreneurs. This study intends to investigate the impact of entrepreneurial orientation (EO) on Industry 4.0 (I4.0) practices of small and medium-sized businesses in light of this scenario and to analyze the mediating role of enviropreneurial marketing (EM) strategy between EO and I4.0 practices. Drawing upon the resource advantage theory and the dynamic capability view (DCV), a framework is developed. A survey approach is adopted to collect data from the owners/managers of 257 electrical and electronics (E&E) small and medium-sized enterprises (SMEs) in Malaysia. To test the study hypotheses, structural equation modeling (SEM) using the partial least squares (PLS) technique is adopted. The findings reveal that enviropreneurial orientation and enviropreneurial marketing strategy have positively influenced Industry 4.0 practices in the SMEs of an emerging economy. Furthermore, the results highlight that enviropreneurial marketing strategy positively transmits the effect of enviropreneurial orientation to I4.0 practices. The study outcomes offer some novel contributions and implications to practitioners and researchers.

K. Ali
Universiti Teknologi Petronas, Seri Iskandar, Perak, Malaysia

A. Osman
Richworks International, Shah Alam, Selangor, Malaysia
e-mail: chairman@richworks.com

T. Kanesan (✉)
Executive Office, Proofreading By A UK PhD, Cyberjaya, Selangor, Malaysia
e-mail: drthava@proofreadingbyphd.com

L. Rui
School of Housing, Building, and Planning, Universiti Sains Malaysia, Gelugor, Penang, Malaysia
e-mail: 806975609rui@student.usm.my

R. A. Alfaisaly · A. S. Alalyani
College of Business Administration, Taif University, Taif, Saudi Arabia

N. Hayat, N. C. Nawi (eds.), *Contemporary Issues in Entrepreneurship and Innovative Technology*, Contributions to Environmental Sciences & Innovative Business Technology, https://doi.org/10.1007/978-3-031-50927-8_3

Keywords Enviropreneurial orientation · Enviropreneurial marketing strategy · Industry 4.0 · Dynamic capability view · Resource advantage theory

1 Introduction

Small and medium-sized businesses (SMEs) play a critical role in today's dynamic business environment in both developed and emerging economies (Ali & Johl, 2022a). Manufacturing SMEs in particular are seen as a driving force for industrial economies (Ali & Johl, 2022b). The integration of digital technology with a company's value chain has become more crucial with the emergence of Industry 4.0 (I4.0), also known as the fourth industrial revolution, for developing a more adaptable and flexible business structure. Industry 4.0 is a strategic initiative to alter all corporate activities through digital transformation, establishing a new ecosystem where technology generates and transmits value to stakeholders. Industry 4.0 is not just about technology; it also includes people and processes. According to Machado et al. (2021), I4.0 is a strategic transformation of all business operations that creates a new ecosystem where technology generates and transfers value to stakeholders.

Although the digital revolution has contributed much to economic progress, it also affects the environment. According to Shift Project (2019), the world's carbon emissions, which have risen by roughly 70% between 2013 and 2020, are attributed to about 3.7% of the growing digitalization. Recently, the idea of environmental entrepreneurship or enviropreneurial orientation, which is strongly tied to digital transformation, has expanded to include many aspects of everyday activities. Recently, the concept of "enviropreneurial orientation" has received increased attention (Ismail, 2022; Khan et al., 2019).

Varadarajan (1992) initially used the term "enviropreneurial orientation" (EO) in a conceptual paper to emphasize the significance of an entrepreneurial mindset in relation to the environment. According to Keogh and Polonsky, the use of both phrases necessitates a commercial focus on innovation and concurrent environmental conservation. Varadarajan (1992) did explain a number of terminologies; however, he did not give a precise explanation of EO. On the other hand, Paulraj (2011) operationalized EO as a firm attribute rather than attitude/motivation or individual factors. Namagembe et al. (2017) define EO based on attitudinal assumptions as a manager's/owner's attitude that combines economic and environmental objectives and encourages the development and implementation of environmentally friendly policies and actions that improve the company's market position and generate revenue. In the same vein, Hartman and Stafford (1998) argue that EO is a type of motivation rather than a strategy. Enviropreneurial orientation (EO) is an entrepreneurial approach that centers around integrating economic, environmental, and social objectives. It is the core of enviropreneurship and aims to solve problems through innovative solutions and technologies, instead of just reacting to government regulations and public pressure. This approach has been described as such by scholars such as Hartman and Stafford (1998) and Namagembe et al. (2017).

Recently, firm stakeholders, particularly consumers, have been concerned with climate degradation and environmental issues. They are more adamant that businesses must contribute to the cause rather than the problem (Randrianasolo, 2021), especially SMEs (Namagembe et al., 2016). However, there are still no effective solutions for dealing with the relevant stakeholders. Enviropreneurial marketing (EM) strategy can be used as a tool to address this problem by reducing stakeholder environmental concerns and promoting I4.0 practices. Environmental issues and entrepreneurial endeavors are incorporated into marketing practices and policies through the use of enviropreneurial marketing strategies (Namagembe et al., 2016; Randrianasolo, 2021).

Despite the fact that prior literature examined the relationship between enviropreneurial orientation, marketing strategy, and firm performance, there is a paucity of empirical evidence in the Industry 4.0 context, especially in SME firms. For instance, Khan et al. (2019) performed an empirical study to investigate the association between EO and small firms' performance through different green marketing strategies. Similarly, Ismail (2022) investigated EO and green marketing to promote sustainability with technology as a moderator in Tanzania. Randrianasolo (2021) examined the role of EM strategy in promoting firm stakeholders in the USA. To the best of the researcher's knowledge, there are not many studies that establish a connection with Industry 4.0 practices despite the fact that there is literature demonstrating that EM attempts to improve entrepreneurial activities pertaining to the environment. Furthermore, EO is still relatively new in the literature (Ismail, 2022). In order to achieve long-term company success, it is essential to understand how SMEs have transformed their entrepreneurial activities to include environmental issues in ordinary business operations. To respond to literature gaps, the current study has the following objectives:

1. To investigate the influence of enviropreneurial orientation on Industry 4.0 practices.
2. To investigate the mediating role of enviropreneurial marketing strategies between enviropreneurial orientation and Industry 4.0 practices.

To achieve the study objectives, a research framework is developed under the theoretical support of the resource advantage theory and the dynamic capability view (DCV) (Khan et al., 2019). The theoretical framework is well explained in the following section. After that, the research methodology and data analysis are applied to attain the research objectives. Finally, a discussion of the study findings and implications is presented in the subsequent section.

2 Theoretical Framework and Hypothesis Development

2.1 *Theoretical Framework*

2.1.1 Resource Advantage Theory (R-A Theory)

The R-A theory emphasizes the importance of capital, knowledge, and innovation in shaping organizational performance as an evolutionary approach to analyzing competitiveness. It asserts that market participants have limited access to information and that entrepreneurship, institutions, and governmental regulations can all have an impact on a firm's capacity to establish and sustain competitive advantages

(Hunt & Morgan, 2017). According to Hunt and Davis (2008), a company's competitive advantage is essential to its survival and success, and businesses should constantly innovate to maintain this advantage.

The R-A theory has contributed to a number of business fields since it first emerged, including strategic management (Thoeni et al., 2016), marketing (D'Andrea, 2020), supply chain management (Hunt & Davis, 2008), and venture creation (Friske & Zachary, 2019). However, there is limited evidence available regarding the application of the R-A theory in entrepreneurship research. For instance, the study of Sahi et al. (2020) was grounded on the R-A theory to examine the relationship between entrepreneurial orientation and SME performance in the I4.0 era. Thus, to address this theoretical gap, the framework of this research is based on the resource advantage theory.

2.1.2 Dynamic Capability View (DCV)

Dynamic capabilities refer to a company's ability to adapt and shape its internal and external resources in response to rapidly changing business environments. This framework, which emerged in the 1990s, suggests that organizations have varying abilities to adapt to changes in their environment (Kessler, 2013). It involves creative and entrepreneurial activity and allows a company to effectively utilize its resources, skills, and assets to meet new market and technological demands by predicting and realigning them accordingly. The taxonomy of DC is primarily based on three primary clusters of competencies: sensing (opportunity identification and assessment), seizing (mobilizing resources to capture the value), and transforming (continual renewal) (Kessler, 2013).

The dynamic capability framework or view is still evolving and has not yet been rigorous. Numerous studies have confirmed the importance of DCV in a specific context. For instance, the work of Ismail (2022) was grounded on DCV to examine the role of green marketing and sustainability in Tanzania SMEs. Likewise, Khan et al. (2019) underpinned the framework on DCV to examine the impact of EO on SME performance in Bangladesh. Finally, Irfan et al. (2019) examined the role of information technology (IT) in promoting SC capabilities and firm agility under the DCV. However, the DCV has been underpinned in a small number of studies, especially in SMEs, within the I4.0 context. Thus, the framework of this research is based on DCV and the R-A theory to fill the theoretical gaps in the literature.

2.2 Hypothesis Development

2.2.1 EO and I4.0 Practices

Enviropreneurial orientation, also known as "ecopreneurship," is recognized as a crucial competency for small and medium-sized enterprises (SMEs) as it expands their responsibilities. In the context of Industry 4.0, enviropreneurial orientation can lead businesses to adopt technologies and practices that are designed to reduce their environmental impact and improve resource efficiency. For instance, businesses that are committed to reducing their carbon footprint may invest in energy-efficient equipment and adopt Internet of Things (IoT) sensors to optimize their energy use. In the same vein, Irfan et al. (2019) argued that enviropreneurial orientation can encourage businesses to experiment with new and innovative approaches to sustainability, such as the use of three-dimensional (3D) printing to produce customized products with minimal waste or the development of circular business models that reduce resource consumption and waste generation. Menon and Menon (1997) argued that EO encompasses a decision-making process that designs and executes actions that are ecologically and entrepreneurially positive, generate profit, and confirm the small firm's business performance goals. This decision-making process, as per the R-A theory and the DCV literature, is essential for SMEs looking to achieve sustainable business outcomes and gain a competitive edge. Additionally, according to the literature, a firm's environmental orientation depends more on its capacity for innovation, proactivity, competition, and risk-taking than it does simply on how well it complies with laws and demands from the public. Furthermore, based on theoretical support, this study argued that to foster EO in SMEs, appropriate levels of capabilities and resources are needed. Carvalho et al. (2018) argued that ecopreneurship promotes manufacturing performance in Industry 4.0. Based on supporting theoretical and literary evidence, the following hypothesis has been put forth:

H1: Enviropreneurial orientation has a positive relationship with Industry 4.0 practices.

2.2.2 Relationship Between EO and EM

A strategy to address environmental issues within marketing, known as "enviropreneurial marketing," places emphasis on being proactive and seeing the environment as one of the company's basic principles (Khattab et al., 2022; Song-Turner & Polonsky, 2016). Because it does not rely on the law or public pressure to push environmental actions, it differs from other environmentally driven approaches. It is a part of corporate social responsibility (CSR), with an emphasis on environmental initiatives and performance (Menon & Menon, 1997; Song-Turner & Polonsky, 2016). Enviropreneurial orientation involves a business mindset that prioritizes sustainability and environmental responsibility, as well as innovation and risk-taking (Khattab et al., 2022). This mindset can inspire businesses to develop marketing strategies that highlight their commitment to sustainability and environmental responsibility (Papadas et al., 2017). For instance, a business with an enviropreneurial orientation may choose to use eco-friendly packaging materials or develop a carbon offset program to reduce its environmental impact. These actions can be communicated to customers through marketing messages that emphasize the business's commitment to sustainability (Nath & Siepong, 2022). EM can also help businesses attract customers who value sustainability and environmental responsibility. By highlighting their environmental initiatives and sustainability efforts, businesses can differentiate themselves from competitors and appeal to consumers who prioritize sustainability in their purchasing decisions (Khan et al., 2019). The enviropreneurial orientation approach is rooted in the R-A theory and the DCV literature. An entrepreneurial focus on the environment can help a small business incorporate environmental principles across the board by influencing its marketing strategy. The following hypothesis is derived from the previous debate:

H2: Enviropreneurial orientation has a positive and significant relationship with enviropreneurial marketing strategy.

2.2.3 Relationship Between EM and I4.0 Practices

In Industry 4.0, the marketing domain is a dynamic business activity. Prior literature indicates that environmental-related marketing strategies help promote Industry 4.0 initiatives. For instance, a business that adopts Industry 4.0 technologies to optimize its energy use and reduce its carbon footprint can use this achievement as a selling point in its marketing (Chou et al., 2022). Kaur et al. (2022) argued that green marketing and I4.0 technologies help reduce the waste in production, brand image and competitive advantage. Likewise, Rosário and Dias (2022) argued that the relationship between traditional environmental-related marketing strategies and I4.0 technologies helps address customer needs more precisely. Based on DCV, Kaur et al. (2022) argued that customer demands are more dynamic in the current digital era; consequently, they required environmental friends' product and services. To satisfy customer demands, the relationship of EM and I4.0 help build a good perception, as a results firm gain dynamic capability and competitive advantage. Thus, based on theoretical and literature support, the following hypothesis has been proposed:

H3: Enviropreneurial marketing strategy has a positive relationship with I4.0 practices.

2.2.4 Mediating Role of EM

The landscape in which international organizations compete has started to change due to environmental concerns. The concept of EM strategy was presented in detail by Menon and Menon (1997). This study highlights the emergence of enviropreneurial marketing and presents this EM strategy as a mediating variable between various organizational factors and firm performance. In the contemporary literature, environmental-related marketing strategy performs a mediating role. EM can mediate the relationship between EO and I 4.0 practices by acting as a catalyst for the adoption of sustainable and advanced manufacturing and production practices (Borah et al., 2022). A business with an EO may be more likely to adopt Industry 4.0 practices that align with their sustainability goals, and by highlighting these practices through their marketing, they can differentiate

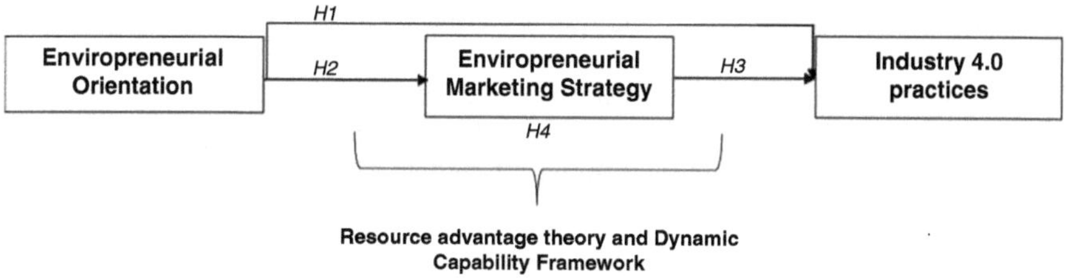

Fig. 1 Theoretical framework

themselves in the marketplace and attract customers who value sustainability (Rainanto et al., 2022). Khan et al. (2019) claimed that environmental-related marketing strategy in Bangladesh mediates the relationship between environmental entrepreneurship and small firm performance. In the case study approach of Song-Turner and Polonsky (2016), they highlight that EM strategy plays a central role in the development of innovative products and services and the creation of new market standards. The following hypothesis has been proposed based on theoretical and literature support:

> H4: Enviropreneurial marketing strategy has a mediating role between enviropreneurial orientation and I4.0 practices.

Finally, Fig. 1 shows the theoretical framework with research hypotheses.

3 Research Methodology

This section discusses the study population and sampling technique in detail. Moreover, this section highlights the operational measurements of variables adapted in this study.

3.1 Population and Sampling

This research focuses on electronic and electrical (E&E) small and medium-sized enterprises (SMEs) in Malaysia. According to the SME Corporation Malaysia (2013), a business with 5–74 employees is considered small and one with 75–200 employees is considered medium. The number of full-time employees was used to categorize the firms into small and medium (Ali & Johl, 2022a). To gather data, an online questionnaire was sent to potential respondents through random sampling. The use of online data collection, as per Tanner (2018) and Granello and Wheaton (2004), is more efficient and cost-effective for researchers to reach the target population. Before distributing the questionnaire, a pretest and pilot study were conducted, and minor changes were made for greater clarity. Ethical considerations were also taken into account throughout the research study (Ali et al., 2022; Zhang et al., 2022).

To conduct the survey, a list of electronic and electrical (E&E) small and medium-sized enterprises (SMEs) was obtained from the Federation of Malaysian Manufacturers (FMM) and served as the sampling frame. More than 770 emails were sent based on the FMM list, and 273 survey questionnaires were collected. Following data cleaning, 257 final responses were utilized for data analysis, resulting in a response rate of 33.78%. Table 1 displays the demographic statistics of the final sample data. SmartPLS software, a second-generation structural equation modeling software program that can model latent variables with minimal prerequisites, was used to analyze the data for this research (Hair, Page, & Brunsveld, 2019). This study employed partial least square (PLS) path modeling analysis, which is a form of structural equation modeling (SEM). SEM is advantageous since it enables the examination of complex relationships within a single model (Hair, Page, & Brunsveld, 2019). Furthermore, the selection of PLS was based on flexibility and less strict assumptions. For instance, PLS can handle nominal, interval, or ratio data and does not require normality. It is also appropriate for small samples (Hair, Page, & Brunsveld, 2019).

3.2 *Measures*

The variables for the study's items have been adapted from previous research. The six items used to test the enviropreneurial orientation (EO) construct were taken from Khan et al. (2019). Eight items were used to measure the enviropreneurial marketing (EM) strategy, which were taken from Randrianasolo (2021). In order to

Table 1 Demographic statistics of the sample ($N = 257$)

"Variable	Item	Frequency	Percentage (%)"
"Gender"	"Male	147	57.20
	Female"	110	42.80
Age	20–25	41	15.95
	26–30	34	13.23
	31–35	52	20.23
	36–40	64	24.90
	41–45	26	10.12
	46–50	23	8.95
	≥51	17	6.62
Education	Intermediate	53	20.62
	Diploma	57	22.18
	Graduation	79	30.74
	Master	41	15.95
	Others	27	10.51
"Work experience"	"≤5 years	77	29.96
	6–10 years	97	37.74
	≥11 years"	83	32.30

measure Industry 4.0 practices, six questions were adapted from Stentoft et al. (2021) and Ali and Johl (2022a) (I4.0). The study items were evaluated using a seven-point Likert scale.

4 Data Analysis

4.1 Descriptive Statistics and Common Method Bias

The mean, kurtosis, and skewness, as well as other study constructs, are presented in this part, along with a descriptive analysis of each. The results of the analysis are provided in detail in Table 2. Except for EM, which has a mean value of 3.591, the table indicates that the mean values of the remaining variables are higher than 4.0. Additionally, the skewness and kurtosis values are within the threshold limit of ±2, as per George and Mallery (2010). The correlation statistics presented in Table 2 indicate that all the study variables have moderate to high levels of relationship. To ensure that the dataset is unbiased, an analysis of common method bias (CMB) was performed. Both procedural and statistical remedies were adopted, as recommended by Podsakoff (2003), to eliminate any CMB issues. The Harman single-factor test was applied statistically. A single factor explained less than 50% of the variance, according to the exploratory factor analysis. Additionally, according to Kock (2015) and Kock and Lynn (2012), the dataset can be regarded as being free from CMB concerns if the variance inflation factor (VIF) score in the full collinearity test is less than or equal to 3.3."

4.2 Assessment of the Measurement Model

4.2.1 Reliability and Convergent Validity

According to Hair, Page, and Brunsveld (2019), "the assessment of convergent validity, which comprises of both reliability and validity, is a part of the assessment of the measurement model. Cronbach's Alpha, composite reliability (CR), and rho A were used to evaluate the data's reliability. Values of Cronbach Alpha, CR, and rho A that are more than or equal to 0.70 are regarded as acceptable (Hair, Page, &

Table 2 Descriptive statistics and correlation

							Correlation		
Constructs	Min	Max	Mean	SD	Skewness	Kurtosis	(1)	(2)	(3)
EO (1)	1	7	4.537	1.836	−0.288	−1.119	1		
EM (2)	1	7	3.591	1.880	0.239	−1.192	0.322	1	
I4.0 (3)	1	7	4.000	1.905	0.085	−1.293	0.512	0.455	1

Brunsveld, 2019; Liu et al., 2015)." In terms of convergent validity, "item loading and average variance extracted (AVE) are critical (Hair, Page, & Brunsveld, 2019). According to Hair, Page, and Brunsveld (2019), item loadings greater than or equal to 0.708 are regarded outstanding, however if loadings are between 0.4 and 0.7 and the AVE is 0.5, researchers can keep the constructs." Additionally, AVE should have a threshold value of ≥0.5. Table 3 presents the reliability and convergent validity results.

4.2.2 Discriminant Validity

The discriminant validity analysis assures that each construct reflects unique and distinctive facts. In discriminant validity analysis, Fornell-Larker and cross-loading procedures are commonly utilized; nevertheless, these techniques are known to exaggerate the existence of discriminant validity. To address this, Hair, Sarstedt, and Ringle (2019) recommend using the heterotrait-monotrait ratio of correlations (HTMT) technique as it is more reliable than other techniques. According to Hair et al. (2017), "the HTMT critical value for distinct variables should be lower than 0.85 and 0.90 for related variables. However, Franke and Sarstedt (2019) propose

Table 3 Reliability and convergent validity

				Reliability			
Dimensions	Items	Loading (0.50–0.85)*	VIF (<5)**	Cronbach's α (≥0.70)**	CR (≥0.70)**	rho_A (≥0.70)**	AVE (≥0.50)**
Enviropreneurial orientation (EO)	EO1	0.747	1.647	0.858	0.849	0.859	0.56
	EO2	0.788	1.905				
	EO3	0.751	1.691				
	EO4	0.769	1.760				
	EO5	0.767	1.761				
	EO6	0.768	1.827				
Enviropreneurial marketing (EM) strategy	EM1	0.696	1.649	0.887	0.91	0.892	0.585
	EM2	0.714	1.627				
	EM3	0.779	1.926				
	EM4	0.748	1.764				
	EM5	0.714	1.696				
	EM6	0.755	1.836				
	EM7	0.722	1.686				
	EM8	0.846	2.541				
Industry 4.0 (I4.0) practices	I1	0.841	2.510	0.926	0.942	0.929	0.731"
	I2	0.861	2.746				
	I3	0.853	2.700				
	I4	0.868	2.774				
	I5	0.849	2.624				
	I6	0.857	2.637				

that the maximum HTMT threshold value for all related and unrelated variables be less than 0.90. The results of the discriminant validity analysis, as shown in Table 4."

4.3 Assessment of the Structural Model (H1–H3)

The association between exogenous and endogenous factors was evaluated in the structural model assessment, as proposed by Hair, Page, and Brunsveld (2019). The structural model findings, *t*-value, confidence interval (CI), and decision criteria are shown in Table 5. Based on the analysis, the association between enviropreneurial orientation (EO) and Industry 4.0 practices (I4.0) was examined." The empirical results indicate that the EO➔I4.0 (H1: $\beta = 0.425$, $t = 6.282$, $p < 0.05$) relationship is positive and significant. "Thus, H1 has been accepted. Furthermore, the empirical results highlight that EO➔EM (H2: β=0.454, t= 8.233, p<0.05) relationship is positive and significant. Therefore, H2 has been accepted. Finally, the results indicate that the relationship of EM➔I4.0 (H3: β=0.368, t= 5.651, p<0.05) is positive and significant. Thus, H3 has been accepted."

4.3.1 Predictive Relevance (Q2) and Coefficient of Determination (R2)

The Q2 value is used by the PLS-SEM approach to assess the model's accuracy and predictive significance. The blindfolding approach, as described by J Hair et al. (2017), is used to calculate the Q2 value. In the case of reflecting models, this method properly and effectively displays the data points of variables. "When the value of Q2 in a structural model is greater than zero, it shows that some exogenous variables are predictive of specific endogenous variables (Nitzl et al., 2016). In terms of predictive relevance, J Hair et al. (2017) describe a Q2 value of less than or

Table 4 Discriminant validity

	Fornell-Larcker criterion			HTMT		
Constructs	1	2	3	1	2	3
EO (1)	0.765					
EM (2)	0.454	0.748		0.515		
I4.0 (3)	0.592	0.561	0.855	0.659	0.612	

Table 5 Structural model

Relation		Direct effect	*t*-value	CI [2.05%–97.5%]	Decision
H1	EO➔I4.0	0.425	6.828	[0.322–0.525]	Supported
H2	EO➔EM	0.454	8.233	[0.355–0.537]	Supported
H3	EM➔I4.0	0.368	5.651	[0.259–0.473]	Supported

equal to 0.02 as weak, less than or equal to 0.15 as medium, and less than or equal to 0.35 as strong." Table 6 shows the Q2 values; the predictive relevance of EM (0.111) is medium, and I4.0 (0.328) falls under the strong category.

Table 6 presents the coefficient of determination (R^2), which illustrates how exogenous variables explain the relationship with endogenous variables. "According to Hair et al. (2017), R^2 values less than or equal to 0.25 indicate a weak association, R^2 values less than or equal to 0.50 indicate a moderate association, and R^2 values less than or equal to 0.75 indicate a strong association. Based on Table 6, EM (0.207) indicates a weak, and I4.0 (0.457) shows a moderate value of R^2."

4.3.2 Mediation Analysis (H4)

A novel technique was used to examine the role of EM as a mediator between EO and I4.0. The direct effect, "as proposed by Carrión et al. (2017) and Nitzl et al. (2016) EO➔I4.0 (path c`) was examined." Table 7 indicates that the relationship between EO and I4.0 is positive and significant. Then the indirect relationships of EO➔EM (path a) and EM➔I4.0 (path b) are examined. Both relationships are positive and significant. As recommended by Hair et al., the strength or magnitude of mediation is necessary for complex path modeling. "It can be measured through variance accounted for (VAF). Hair et al. recommended that a VAF value less than 0.20 represent no mediation, $0.20 \leq \text{VAF} \leq 0.80$ represent partial mediation and greater than 0.80 represent full mediation. Table 7 shows that EM has a partial mediation between EO and I4.0 because the value of VAF is greater than 0.20." Furthermore, Fig. 2 shows the complete PLS path model.

4.3.3 Importance-Performance Map Analysis (IPMA)

An importance-performance map analysis (IPMA) is employed in addition to the PLS-SEM results to present the results in a four-quadrant diagram, as illustrated in Fig. 3. The vertical axis reflects variable performance, whereas the horizontal axis denotes construct relevance. According to Ringle and Sarstedt (2016), "when

Table 6 Q2 and R2 statistics

Variables	R2	Q2
EM	0.207	0.111
I4.0	0.457	0.328

Table 7 Mediation analysis

Path		"Relationship	β-value	*p*-value	VAF"
Direct effect	Path c`	EO➔I4.0	0.592	0.001	22%
Indirect effect	Path a	EO➔EM	0.454	0.001	
	Path b	EM➔I4.0	0.368	0.001	

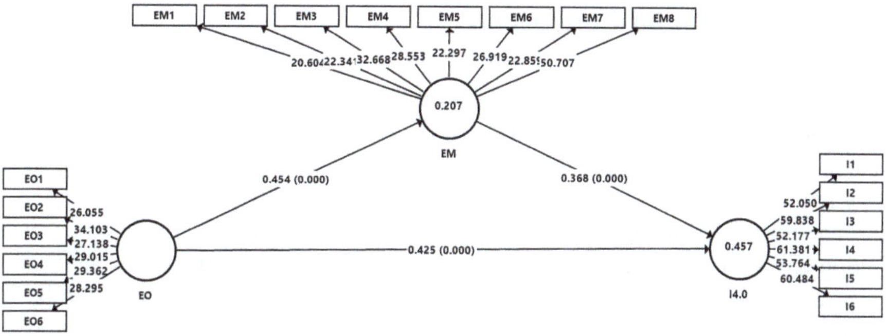

Fig. 2 PLS path model

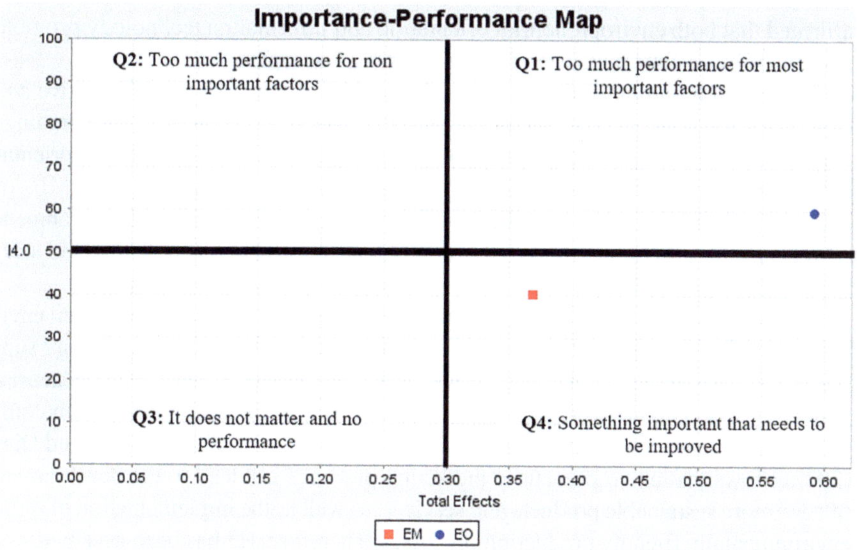

Fig. 3 Important-performance map analysis

analysing the IPMA, the attributes that fall in the lower right area (Q4) are of higher interest for improvement, followed by the upper right (Q1), lower left (Q3), and finally, the upper left (Q2). Therefore, the IPMA guides prioritizing managerial activities that are important for achieving the selected target but require performance improvement." Fig. 3 illustrates that EM is the most important construct for maximizing the performance of Industry 4.0; hence, it falls under Q4. On the other hand, EO is a less important construct but performs well in achieving Industry 4.0, thus it falls under Q1.

5 Discussion and Conclusion

This study aims to empirically examine the role of enviropreneurial orientation in promoting Industry 4.0 practices in the SMEs of emerging economies. Furthermore, the investigation emphasizes the mediating role of enviropreneurial marketing strategy in the link between enviropreneurial orientation and IR4.0 practices.

The empirical findings of the first objective highlight that enviropreneurial orientation has a positive and significant effect on IR4.0 practices and enviropreneurial marketing strategies. The finding indicates that enviropreneurial orientation has a positive and significant relationship with I4.0 practices. The results are in line with past studies. The empirical work of Khattab et al. (2022) indicated a significant relationship between enviropreneurial orientation and Industry 4.0 practices (big data analytics) in emerging economies. Furthermore, the work of Ismail (2022) affirmed that both enviropreneurial orientation and information technology promote sustainability in Tanzania. Therefore, H1 has been accepted.

The research findings indicate that enviropreneurial orientation has a positive and significant relationship with enviropreneurial marketing strategies. These findings corroborate the ideas of Khan et al. (2019), who argued that enviropreneurial orientation positively promotes environmental-related marketing strategies in the SMEs of emerging economies. Moreover, Rasi et al. argued that entrepreneurial environmental-related activities help promote enviropreneurial value chains (marketing and sale) in Malaysian SMEs. Based on the literature support, H2 has been accepted.

The empirical results indicate a positive and significant association between enviropreneurial marketing strategy and Industry 4.0 practices. The study results are supported by past studies. A study by Kaur et al. (2022) shows that enviropreneurial marketing strategies can also help companies differentiate themselves from their competitors and create new business opportunities. Likewise, a study by Rosário and Dias (2022) shows that enviropreneurial marketing practices can lead to the development of new, more sustainable products and services, as well as the implementation of more environmentally friendly production processes. Therefore, H3 has been accepted.

Finally, the second research objective indicates that enviropreneurial marketing strategy has a mediating role between enviropreneurial orientation and Industry 4.0 practices. This finding broadly supports the work of other studies in this area. For instance, Menon and Menon (1997) argued that enviropreneurial marketing strategy has a mediating relationship between exogenous and endogenous variables. In the same vein, Khan et al. (2019) confirmed the mediating role of environmental marketing strategy between enviropreneurial orientation and SMEs' performance. Thus, H4 has been supported.

5.1 Theoretical Contributions

The research findings offer threefold novel theoretical contributions. First, this study contributes to the existing understanding of the resource-advantage theory by exploring its application in addressing environmental challenges in the digital age.

The results indicate that an enviropreneurial orientation and marketing strategy can provide SMEs with a distinct competitive advantage, particularly in adopting Industry 4.0 practices. Second, the research outcomes extend the pertinent literature by appropriating environmental strategic orientations with Industry 4.0 practices. Prior studies examined EO with other outcome variables, but authors failed to find any study that establishes a link between EO, EM, and I4.0. Thus, this study advances the past literature by developing a relationship among study variables. Finally, this study contributes to the integration of the resource-advantage theory and the dynamic capability perspective. This study highlights that EM and EO are not just a resource but also dynamic capabilities and precursors to I4.0 practices. Thus, this study contributes to the DCV literature by constructing the structural relationship between EM, EO, and I4.0 practices.

5.2 *Practical Implications*

Apart from theoretical contributions, the study findings also offer pragmatic implications to managers and practitioners. First, the study affirms that EO and EM promote Industry 4.0 practices in SMEs. Thus, owner-managers should cultivate more environmental-related strategies and carefully apply them to build the proper way of timely implementing I4.0. Second, SME owner-managers may have proper plans for introducing a visible digital process. This must be in line with enviropreneurial orientation and marketing strategies. Furthermore, it is important that SME owner-managers design proper on- and off-job training so that employees may be able to demonstrate environmental-related strategies. Finally, as stated in prior studies, most of the SMEs in emerging countries have poor adoption of I4.0 technologies. Thus, owner-managers must continuously create a learning culture so that the ability of the workforce may be enhanced to adopt new and innovative technologies.

5.3 *Limitations and Future Research Avenues*

Although the research was performed with care and thoroughness, there are some limitations that suggest future research avenues. Firstly, as the research is focused exclusively on SMEs, it is possible that the findings may not fully apply to large firms with different structures, resources, and objectives. Therefore, it is important to acknowledge this limitation and conduct additional studies to assess whether similar outcomes are observed in larger firms. This would enhance the generalizability of the results and provide a more comprehensive understanding of the phenomena under investigation. Secondly, the current study solely focused on the impact of enviropreneurial orientation (EO) and enviropreneurial marketing (EM) factors on promoting I4.0. Therefore, it is recommended for future research to explore the involvement of other variables, both continuous and discrete. Additionally, it would be intriguing to investigate the intricate relationships among various mediators and moderators, which could provide a more comprehensive understanding of the underlying mechanisms that drive the adoption of I4.0 in organizations. Thirdly and lastly, in future investigations, the longitudinal method can be employed to examine how the variables examined in this study evolve and transform over an extended period of time.

References

Ali, K., & Johl, S. K. (2022a). Impact of total quality management on industry 4.0 readiness and practices: Does firm size matter? *International Journal of Computer Integrated Manufacturing, 36*, 1–23.

Ali, K., & Johl, S. K. (2022b). Entrepreneurial competencies needed to improve innovation performance in industry 4.0: An artificial neural network approach. In *Entrepreneurial innovations, models, and implementation strategies for industry 4.0* (pp. 1–19). CRC Press.

Ali, K., Johl, S. K., Muneer, A., Alwadain, A., & Ali, R. F. (2022). Soft and hard total quality management practices promote industry 4.0 readiness: A SEM-neural network approach. *Sustainability, 14*(19), 11917.

Borah, P. S., Pomegbe, W. W. K., & Dogbe, C. S. K. (2022). Mediating role of green marketing orientation in stakeholder risk and new product success relationship among European multinational enterprises in Ghana. *Society and Business Review, 17*(4), 485–505.

Carrión, G. C., Nitzl, C., & Roldán, J. L. (2017). Mediation analyses in partial least squares structural equation modeling: Guidelines and empirical examples. In *Partial least squares path modeling: Basic concepts, methodological issues and applications* (pp. 173–195).

Carvalho, N., Chaim, O., Cazarini, E., & Gerolamo, M. (2018). Manufacturing in the fourth industrial revolution: A positive prospect in sustainable manufacturing. *Procedia Manufacturing, 21*, 671–678.

Chou, S. F., Horng, J. S., Liu, C. H., Yu, T. Y., & Kuo, Y. T. (2022). Identifying the critical factors for sustainable marketing in the catering: The influence of big data applications, marketing innovation, and technology acceptance model factors. *Journal of Hospitality and Tourism Management, 51*, 11–21.

D'Andrea, F. A. M. C. (2020). Strategic marketing & Austrian economics: The foundations of resource-advantage theory. *The Review of Austrian Economics, 33*(4), 481–501.

Franke, G., & Sarstedt, M. (2019). Heuristics versus statistics in discriminant validity testing: A comparison of four procedures. *Internet Research, 29*(3), 430–447.

Friske, W. M., & Zachary, M. A. (2019). Regulation, new venture creation, and resource-advantage theory: An analysis of the US brewing industry. *Entrepreneurship Theory and Practice, 43*(5), 999–1017.

George, D., & Mallery, P. (2010). *SPSS for windows step by step: A simple guide and reference 17.0 update* (10th ed.). Pearson.

Granello, D. H., & Wheaton, J. E. (2004). Online data collection: Strategies for research. *Journal of Counseling & Development, 82*(4), 387–393.

Hair, J., Hollingsworth, C. L., Randolph, A. B., & Chong, A. Y. L. (2017). An updated and expanded assessment of PLS-SEM in information systems research. *Industrial Management & Data Systems, 117*(3), 442–458. https://doi.org/10.1108/IMDS-04-2016-0130

Hair, J. F., Page, M., & Brunsveld, N. (2019). *Essentials of business research methods*. Routledge.

Hair, J. F., Sarstedt, M., & Ringle, C. M. (2019). Rethinking some of the rethinking of partial least squares. *European Journal of Marketing, 53*(4), 566–584. https://doi.org/10.1108/EJM-10-2018-0665

Hartman, C. L., & Stafford, E. R. (1998). Crafting 'enviropreneurial' value chain strategies through green alliances. *Business Horizons, 41*(2), 62–73.

Hunt, S. D., & Davis, D. F. (2008). Grounding supply chain management in resource-advantage theory. *Journal of Supply Chain Management, 44*(1), 10–21.

Hunt, S. D., & Morgan, R. M. (2017). The resource-advantage theory of competition: A review. *Review of Marketing Research*, 153–205.

Irfan, M., Wang, M., & Akhtar, N. (2019). Impact of IT capabilities on supply chain capabilities and organizational agility: A dynamic capability view. *Operations Management Research, 12*, 113–128.

Ismail, I. J. (2022). The role of technological absorption capacity, enviropreneurial orientation, and green marketing in enhancing business' sustainability: Evidence from fast-moving consumer goods in Tanzania. *Technological Sustainability, 2*(2), 121–141.

Kaur, R., Singh, R., Gehlot, A., Priyadarshi, N., & Twala, B. (2022). Marketing strategies 4.0: Recent trends and technologies in marketing. *Sustainability, 14*(24), 16356.

Kessler, E. H. (Ed.). (2013). *Encyclopedia of management theory*. Sage Publications.

Khan, E. A., Royhan, P., Rahman, M. A., Rahman, M. M., & Mostafa, A. (2019). The impact of enviropreneurial orientation on small firms' business performance: The mediation of green marketing mix and eco-labeling strategies. *Sustainability, 12*(1), 221.

Khattab, S., Shaar, I., Alkaied, R., & Qutaishat, F. (2022). The relationship between big data analytics and green supply chain management by looking at the role of environmental orientation: Evidence from emerging economy. *Uncertain Supply Chain Management, 10*(2), 303–314.

Kock, N. (2015). Common method bias in PLS-SEM: A full collinearity assessment approach. *International Journal of e-Collaboration (IJEC), 11*(4), 1–10.

Kock, N., & Lynn, G. (2012). Lateral collinearity and misleading results in variance-based SEM: An illustration and recommendations. *Journal of the Association for Information Systems, 13*(7).

Liu, H., Bahron, A., & Bagul, A. H. B. P. (2015). Justifying scale type for a latent variable: Formative or reflective? *AIP Conference Proceedings, 1691*(1), 050009. https://doi.org/10.1063/1.4937091

Machado, C. G., Winroth, M., Almström, P., Ericson Öberg, A., Kurdve, M., & AlMashalah, S. (2021). Digital organisational readiness: Experiences from manufacturing companies. *Journal of Manufacturing Technology Management, 32*(9), 167–182.

Menon, A., & Menon, A. (1997). Enviropreneurial marketing strategy: The emergence of corporate environmentalism as market strategy. *Journal of Marketing, 61*(1), 51–67.

Namagembe, S., Ryan, S., & Sridham, R. (2017). Enviropreneurial orientation in SME supply chains: Construct measurement development. *World Journal of Entrepreneurship, Management and Sustainable Development, 13*(2), 128–150.

Namagembe, S., Sridharan, R., & Ryan, S. (2016). Green supply chain management practice adoption in Ugandan SME manufacturing firms: The role of enviropreneurial orientation. *World Journal of Science, Technology and Sustainable Development, 13*(3), 154–173.

Nath, P., & Siepong, A. (2022). Green marketing capability: A configuration approach towards sustainable development. *Journal of Cleaner Production, 354*, 131727.

Nitzl, C., Roldan, J. L., & Cepeda, G. (2016). Mediation analysis in partial least squares path modeling: Helping researchers discuss more sophisticated models. *Industrial Management & Data Systems, 116*(9), 1849–1864.

Papadas, K. K., Avlonitis, G. J., & Carrigan, M. (2017). Green marketing orientation: Conceptualization, scale development and validation. *Journal of Business Research, 80*, 236–246.

Paulraj, A. (2011). Understanding the relationships between internal resources and capabilities, sustainable supply management and organizational sustainability. *Journal of Supply Chain Management, 47*(1), 19–37.

Podsakoff, N. (2003). Common method biases in behavioral research: A critical review of the literature and recommended remedies. *Journal of Applied Psychology, 885*(879), 1037.

Rainanto, B. H., Bon, A. T., & Purba, J. H. V. (2022). Environmental management system and pro-environmental behavior in realizing sustainable industry performance: Mediating role of green

marketing management. *International Journal of Global Optimization and Its Application, 1*(1), 12–21.
Randrianasolo, A. (2021). Employing enviropreneurial marketing strategies to gain legitimacy. *Journal of Global Marketing, 34*(2), 131–145.
Ringle, C. M., & Sarstedt, M. (2016). Gain more insight from your PLS-SEM results: The importance-performance map analysis. *Industrial Management & Data Systems, 116*(9), 1865–1886.
Rosário, A. T., & Dias, J. C. (2022). Industry 4.0 and marketing: Towards an integrated future research agenda. *Journal of Sensor and Actuator Networks, 11*(3), 30.
Sahi, G. K., Gupta, M. C., & Cheng, T. C. E. (2020). The effects of strategic orientation on operational ambidexterity: A study of Indian SMEs in the industry 4.0 era. *International Journal of Production Economics, 220*, 107395.
Shift Project. (2019). LEAN ICT. Towards digital sobreity. Available online at: https://theshiftproject.org/wpcontent/uploads/2019/03/Lean-ICT-Report_TheShift-Project_2019.pdf
SME Corporation Malaysia. (2013). Guideline for New SME definition. https://www.smecorp.gov.my/images/pdf/Guideline_for_New_SME_Definition_7Jan2014.pdf
Song-Turner, H., & Polonsky, M. (2016). Enviropreneurial marketing in greening corporate activities: Evidence from four Chinese green firms. *European Business Review, 28*(5), 506–531.
Stentoft, J., Adsbøll Wickstrøm, K., Philipsen, K., & Haug, A. (2021). Drivers and barriers for industry 4.0 readiness and practice: Empirical evidence from small and medium-sized manufacturers. *Production Planning & Control, 32*(10), 811–828. https://doi.org/10.1080/09537287.2020.1768318
Tanner, K. (2018). Chapter 6 - survey designs. In K. Williamson & G. Johanson (Eds.), *Research methods* (2nd ed., pp. 159–192). Chandos Publishing. https://doi.org/10.1016/B978-0-08-102220-7.00006-6
Thoeni, A. T., Marshall, G. W., & Campbell, S. M. (2016). A resource-advantage theory typology of strategic segmentation. *European Journal of Marketing, 50*(12), 2192–2215.
Varadarajan, P. R. (1992). Marketing's contribution to strategy: The view from a different looking glass. *Journal of the Academy of Marketing Science, 20*(4), 335–343.
Zhang, B. S., Ali, K., & Kanesan, T. (2022). A model of extended technology acceptance for behavioral intention toward EVs with gender as a moderator. *Frontiers in Psychology, 13*, 1080414.

Kashif Ali is an accomplished academic and researcher in the field of Management, currently serving as an Assistant Professor at the University of Central Punjab (UCP), Pakistan. With a profound focus on sustainability, quality management, entrepreneurship, and the circular economy, Dr. Ali has made significant contributions to the academic community. His research portfolio reflects his expertise in Total Quality Management (TQM) and its implications for contemporary industry practices, especially in the context of Industry 4.0. Dr. Kashif Ali's diverse research has been published in esteemed journals such as Total Quality Management and Business Excellence, International Journal of Computer Integrated Manufacturing, and Journal of Manufacturing Technology, among others. His expertise is underscored by his engagement as a reviewer for peer-reviewed journals and an academic editor for PLOS One.

Azizan Osman, a distinguished Professor of Practice at Asia e University, Malaysia, and a Business Growth Mentor, possesses extensive expertise in business development and marketing strategies. His unwavering commitment to lifelong learning propels his entrepreneurial journey. Holding a Ph.D. in Business (Asia e University) and an MSc in Marketing (Salford), he is an alumnus of esteemed institutions, including Columbia Business School, Kellogg School of Management, and IMD Business School. He is a Chartered Management Accountant (CIMA, UK) and a Fellow of the Chartered Institute of Marketing (FCIM, UK).

Azizan leads as the Chairman of RichWorks International and holds a significant stake in the publicly listed company ManagePay Systems Berhad (MPay), recently being appointed as a Board

Member. As a Business Growth and Marketing Strategist, he guides entrepreneurs through events and programs, offering mentorship and inspiration through academia and practical guidance. His profile epitomizes lifelong learning, leadership, and entrepreneurship, motivating business professionals across academic and practical realms.

Thavamaran Kanesan completed his Ph.D. degree in Engineering in less than 3 years, majoring in the integration of 4th Generation Long-Term Evolution (LTE) and optical communications. Thereafter, he solely founded Proofreading By A UK Ph.D., an Academic Consulting Firm that provides services across many verticals. The firm has grown from having a single founder to currently employing over 102 Consultants all over the globe and currently serves Corporates/ Universities in over 34 countries, including UK, Saudi Arabia, Malaysia, Korea, Australia, and many more. In his personal Research Journey, Thavamaran has an exciting publication record and awards, with more than 60 publications (list attached), mostly published in (SCI/SSCI–ISI) Quartile 1 journals and Low Acceptance Rate conferences. Most notably, he had garnered over +105,000 Academic Followers in his vibrant Facebook page/group, which completely focuses on Academic and Career Paradigms. Proofreading By A UK Ph.D. currently serves as the official editorial for many universities, Governmental and Private Organizations, notably MABMS, MARDI, USM GBS, UniMAS JCEST, UMK JEB, USIM, UPM Civil, UKM Social Science, and OZZIE AMDHS. Apart from his entrepreneurship journey, he was also with the prestigious Aston Institute of Photonic Technologies, Birmingham (UK) as a postdoctoral researcher and worked in European FP7 project FOX-C and EPSRC UNLOC focusing on all optical-orthogonal frequency division multiplexing (AO-OFDM) and coherent O-OFDM (CO-OFDM). In order to gain further industrial insights, he joined Telekom Malaysia as a Technical Manager, focusing on radio-over-fiber system projects and directly managed 24 members of technical staffs. To date, Thavamaran has managed to secure over RM 23.7 Mil (USD 5.4 Mil) worth of Research Grants and won numerous Awards, namely BPH Successful People in Malaysia, Best Papers, and Researcher-of-the-year awards..

Liang Rui is a Ph.D. student at the School of Housing, Building and Planning, Universiti Sains Malaysia. She got a Master's degree from ShanDong University of Arts, China. Her research focuses on Design Management, such as Interior Design.

Reem Abdulrahman Alfaisaly is an Assistant Professor of Marketing at the College of Business Administration, Taif University, Kingdom of Saudi Arabia. She has more than seven years of experience as an academician. She holds a DBA in Marketing from the National University of Malaysia, Graduate School of Business (UKM-GSB) in 2018 and specializes in Marketing. Her research interests are in the areas of Tourism and Hospitality Marketing, Electronic Marketing, Consumer Behavior, Strategic Marketing Entrepreneurships as well as Services Marketing. Her research publications have appeared in the International Journal of Hospitality and Tourism Systems. Reem Alfaisaly is also the co-author of "E-Marketing (2nd ed) 2019," an academic book published by ETRAC for printing publishing and distribution. Reem is the recipient of numerous teaching and research awards at both the faculty and university levels. She is passionate about knowledge sharing and believes that real projects or practical applications should be embedded in the teaching process..

Abdulrahman Saleh Alalyani is an Assistant Professor of Marketing at the College of Business Administration, Taif University, Kingdom of Saudi Arabia. He has more than six years of experience as an academician. He holds a DBA in Marketing from the National University of Malaysia, Graduate School of Business (UKM-GSB) in 2019 and specializes in Marketing. His research interests are in the areas of Tourism and Hospitality Marketing, Electronic Marketing, Consumer Behavior, Strategic Management, Strategic Marketing Entrepreneurships as well as Services Marketing. His research publications have appeared in the International Journal of Hospitality and Tourism Systems. His teaching performance has consistently been ranked among the highest, and in recognition of his achievement at the faculty and university.

Conceptual Framework on Local Knowledge Practices with the Implementation of Branding and Packaging for Fermented Salted Fish (*Ikan Bekok*) to Mobilize Human Capital Empowerment

Tengku Fauzan Tengku Anuar, Nurul Syahida Mat Hussin, and Ren Chen Yang

Abstract Branding and packaging is a growing trend among consumers to purchase local products. However, it is a constraint toward preceding fishery activity in empowering human capital which confine to traditional ways of handling and only adequate for daily necessity. Therefore, this study aims to (1) explore the potential of branding and packaging in the operational chain scope for *ikan bekok* products, and (2) develop a conceptual framework of branding and packaging potential incorporating with social enterprise implementation, human capital empowerment, and Sustainable Development Goals (SDGs). This study involves the methods of reviewing prior literature, module delivery, observation, semi-structured interview, and proposing innovation on products branding and packaging. This fermented salted fish project has been implemented in Tumpat, Kelantan, Malaysia. This research has developed a conceptual framework that links branding and packaging with social enterprise implementation, human capital empowerment, and the Sustainable Development Goals (SDGs) related to quality education (SDG 4) and decent work and economic growth (SDG 8). The study acknowledges the significance of local knowledge in sustaining the livelihoods of the coastal community, as well as its potential for poverty reduction and improving living standards. Overall, this research contributes to the literature on sustainable development by emphasizing the mer-

T. F. T. Anuar (✉) · N. S. M. Hussin
Faculty of Creative Technology and Heritage, Institute for Poverty Research and Management (InsPek), Universiti Malaysia Kelantan, Kota Bharu, Kelantan, Malaysia
e-mail: tengkufauzan@umk.edu.my; nurulsyahidamathussin@umk.edu.my

R. C. Yang
Faculty of Creative Technology and Heritage, Universiti Malaysia Kelantan, Kota Bharu, Kelantan, Malaysia
e-mail: C20D0204F@siswa.umk.edu.my

N. Hayat, N. C. Nawi (eds.), *Contemporary Issues in Entrepreneurship and Innovative Technology*, Contributions to Environmental Sciences & Innovative Business Technology, https://doi.org/10.1007/978-3-031-50927-8_4

gence and mobilization of human capital empowerment with the approach of branding and packaging for local products.

Keywords Conceptual framework · Local knowledge · Branding · Packaging · Human capital empowerment

1 Introduction

Ikan bekok or fermented salted fish is one of the posts—harvest handling of catch products produced by the coastal community in the East Coast of Peninsular Malaysia. The fish utilization activities are part of the coastal community's local gastronomic knowledge and practices, which involve curing the perishable products for sustaining and more extended shelf–life storage (Haji Megat Abu Bakar, 2022; Hua, 1989). In respect of human capital empowerment, local knowledge practice among the coastal community frequently discussed poverty-related (Rhoumah, 2016; Nuridayati & Hasan, 2021), consumerism related (Davis & Halligan, 2002), and factors of improving the living standards of the community. Therefore, taking the coastal community activity in making fermented salted fish at Tumpat, Kelantan as the case study, this study aims to elaborate the conceptual framework on how branding and packaging related with the implementation of social enterprise, empowering human capital and at once sustaining the local knowledge as it complies with the Sustainable Development Goals (SDGs) referring to quality education (SDGs 4) and decent work and economic growth (SDGs 8).

It is undeniable that local knowledge has the potential for human capital empowerment. The organizational system in the fisheries curing process includes the production scope and job chain, leading to today's diversion of technological development. Meanwhile, the fisheries activities preceding comes with the confine to traditional methods and it just adequate for daily necessities, then later it is to meet the demand of the mainstream society. Therefore, human capital empowerment with the emergence of local knowledge significantly becomes the sustainable development agenda due to globalization and market job saturation issues followed by the decline of economic activity in various sectors.

2 Literature Review

2.1 Human Capital Empowerment in Coastal Community

At a glance, a usable productive resource is implied in the term "capital." It also can be defined by the following four types proposed by Smith (2007): land, buildings, machinery, and human abilities. Besides, "capital" also broadly includes *all the forms of assets and capabilities—natural, biological, financial, and human—that can be harnessed for human development* (Šlaus & Jacobs, 2011). Therefore, this paper refers to "human capital" as a social asset that encompasses training, education, and professional initiative to improve the level of knowledge, skills, and abilities. Human capital becomes the determinant for long term sustainability, and parallel piecemeal contribute to economic growth locally and globally (Marimuthu et al., 2009).

There are five (5) factors of influence identified stimulating human capital to enhance economically from the previous studies—(1) education, (2) ownership and use of new human capital, (3) communicative actions, (4) corporate social responsibilities (CSR), and (5) local government involvement as summarized in Table 1. These factors were found to influence social development internally and externally. Firstly, for internal influence, education has been regarded as the *foundation of human development* (Mei et al., 2019). Therefore, it is not limited to just formal or nonformal educational system approaches for yielding better economic and social returns (Sembok & Ismail, 2021). In addition, the educational reinforcement toward the human capital be ensured by giving the appropriate skills by training and upgrading the teaching quality (Ejere, 2011; Huffman & Quigley, 2002).

Instead of educational, ownership toward the new human capital and communicative actions can also be referred to as the internal factors in stimulating human capital. The previous study on the transitional change of traditional activities among Orang Kuala to ownership of new human capital shows that they tend to recover the economic situation and get out from the poverty trap (Zal, 2008). In addition, as accelerating their consolidation among the coastal community, communicative action become process of self-help. It is the most effective forum to adopt and adapt the new digital approach which parallelly move them to live in the existence condition (Mirodiyatun Resi Nuridayati & Hasan, 2021). The undeniable external factors to empower the human capital leads to get the acquisition help from government, local government and non—governmental organizations (Rajab et al., 2022; Rahmawati & Apip, 2022). These agencies come with the intertwine strategies by giving the guidance, training, management, and distribution to communities to increase the quality and productions. Therefore, human capital can be generated from existing communities' local knowledge and can be improvised as it appropriate with the necessities for better social and economic development.

Human capital among the coastal community usually related to the activities based on harvesting a catch from sea. It uses as daily food need for family and neighborhood market, then expanded to small and medium scale fishing enterprise (Zamroni et al., 2017). It uses as daily food needs for family for the first place then expanded as neighborhood market. Like another coastal community, Tumpat residency located in *Pantai Sri Tujuh*, they also own human capital. Sea resources based activities by the coastal community involved (1) fishermen—who are mostly dominated by men and actively catching fish, (2) fisheries enterprise—who are mostly played the role by the fisher's wives and running the small and medium scale fisheries products (Frangoudes & Keromnes, 2008). Moreover, in order to empower herself, women in developing economies creatively carved out niches job for themselves

Table 1 Factors empowering human capital and related role player in the local communities

Authors	Title	Region	Community/ area	Factors empowering human capital	Role player
Huffman and Quigley (2002)	The Role of the University in Attracting High Tech Entrepreneurship: A Silicon Valley Tale	San Francisco	Silicon Valley	– Education – Entrepreneurial talent	Academic institution
Zal (2008)	The Ownership and Use of New Human Capital Among Orang Kuala in Malaysia and Its Relation with Economic Empowerment	Malaysia	Orang Kuala	– Ownership and the use of new human capital	Business-based economic activities
Ejere (2011)	Human Capital Formation as Catalyst for National Development	Nigeria	General	– Education – Quality of teachers – Training (formal and informal training) – Health and nutrition	Educational system
Mirodiyatun Resi Nuridayati and Hasan (2021)	Empowerment Communications as a New Perspective in the Empowerment of Coastal Communities of Pase Raya	Indonesia	Pase Raya	– Communicative actions	Social groups
Rahmawati and Apip (2022)	Partnership Model for the Implementation of Corporate Social Responsibility States Own Company in Coastal Community Empowerment Programs in Cilegon City	Indonesia	Cilegon City	– Corporate social responsibility (CSR)	Nongovernmental organization (NGO)

(continued)

Table 1 (continued)

Authors	Title	Region	Community/ area	Factors empowering human capital	Role player
Rajab et al. (2022)	The Role of the Marine and Fisheries Office in Empowering Fishermen in Majene Regency	Indonesia	Majene Regency	– Local government assistance	Local government

(Boateng, 2023). In this study, human capital can be seen through the activities and lengthy process of handling, curing, and even processing the fisheries products.

As appropriate with the market demand, some of fishery's products has been categorized as perishable products which potentially can penetrate the higher market segmentation. However, complaining toward the hygienic, shelf—lifetime consuming and unpleasant smell lead to acknowledging the power of brand and packaging. Therefore, the expansion chain of handling fishery's products can undergo positive change with filling the gap by improving and empowering the brand and packaging knowledge among the fisheries entrepreneurs.

2.2 *Local Knowledge Practices and Social Enterprises*

Local knowledge practices have eventually been established through social and cultural processes. It may also define as one of the identities that can be represent the community in certain geographical areas. The rural societies and indigenous practice deal with the ecological complexity and dynamic significantly would perceive its continuum from nature then shape subsistence activities (Peloquin, 2009). Although the findings among the coastal Bangladesh shows formal institutions excluding indigenous and local knowledge from social learning process (Choudhary et al., 2021), it is ascertain that utilizing the aquatic resources prompt to initiate the economic development and conservations toward environment (Vayda et al., 2019; Aulia et al., 2020; Ruddle, 1994). Therefore, it is undeniable for operational "folk-man-ages system" revitalization which tend to build the community resilience and inner strength outcomes by their own way.

The wide recognition on the role of local knowledge in promising human capital outcomes has given the impact and changes toward the communities' subsistence activities. Encourage the local knowledge activities tend to offer the employment skills, experience-based and maintaining the business activities (Steiner et al., 2019). The creation of the community-based business activity can represent as an alternative mechanism for programming the traditional rural development (Berkes & Davidson-Hunt, 2007; Maguirre et al., 2017). The interest toward the social enterprise study in remote area has grown exponentially when it is integrating the social values, the financial returns, and addressing the environmental issues (Douglas et al., 2018; Steiner et al., 2019). Therefore, this study keen into the relevance of social enterprise which is highly recommended to get sustain the local knowledge development, increase the income revenue, upgrading skills, and might provide employment opportunities for surrounded population.

Social enterprise of local communities' activity can be influenced by its demographic, ethnic, and cultural that will claim as economic and politics tension (Douglas et al., 2018). Emphasizing the definition on social enterprise, it refers to nonprofit operational business, yet become the raise of the communities' revenue, upgrading skills with social value (Zainon et al., 2014). This social enterprise become a means tool for negotiating with the processes of globalization and the commercial goals (Berkes & Davidson-Hunt, 2007; Cheah Sau Seng, 2018) which can lead to solve the societal problem, achieve social and ecological change. Therefore, this study can relate between local knowledge practices and social enterprise that happened among the coastal community in Tumpat Kelantan who already incorporate unconscious and undoubted to these practice and enterprise.

It is common to acknowledge the economic activities among the coastal population is fisheries—based products. Tumpat residency who stay in coastal area averagely handling, curing, and produce their post—harvest fisheries into products like crackers, salted dried fish, shrimp, and squids, fermented salted fish or 'Iikan bekok," fermented anchovies (*budu*), and fermented fish (*pekasam*). Instead of retail selling on these ready to cook products, some practitioners step ahead to serve ready to eat products as an exclusive gastronomic delight by offering the locals organoleptic cuisine. Taking the gaining popularity product of *ikan bekok*, there are lengthy of process involved in preparing *ikan bekok* which the practices still confined to the traditional technology. Yet, the chain process implied benefits to single mothers and locals to gain knowledge, experiences, skills, and job. Therefore, social enterprise seem has been embedded among the practitioners who cultivate *ikan bekok* by *employing low-skilled people to provide job experiences* (Steiner et al., 2019). This is enterprising voluntary charities by the practitioners that aware a community need and want to consider as self-help and helping others.

2.3 Branding and Packaging in Empowering Human Capital

Branding is not only capable for attracting necessities products but has been expanded as the "nation branding" (Saad, 2020). It indicates that the branding significantly accepted by the holistic operational as positive perception in offering products. Besides that, brand equity and human capital also become the critical asset for short—term and long—term stability and success (Kim & Sturman, 2012). Meanwhile, in the intervening period of time, local knowledge and the competitive willingness has shown and facilitate the sustainable development in countryside by combining their specialization, fragmentation, and build the quality strategy (Marsden & Smith, 2005). Therefore, this research tends to put the influence of the brand and the action toward its packaging for scaling up the managerial quality by providing the additional chain of empowering human capital among the coastal community with branding and packaging.

In glance, "brand" and "branding" are bandied term that numerously discussed to prompt the effect and power of brand. It is because *two-third of consumer purchased products are influenced by brand* (Davis & Halligan, 2002). Rossiter (2014) has given detail definition on the brand by emphasizing the influence of brand awareness and brand attitude, while branding involved with the cognitive and emotive bond. In simple words, *a product is what you sell, a brand is the perceived image of the products you sell, and branding is the strategy to create that image* (Marion, 2021). Meanwhile, parallel to the aforementioned, packaging is an inseparable element in forming the images and brand (Ardhianto & Hardjana, 2016). It is a trend fostering growth for convenience, well functionate, and indulgence toward the developed packaged food (Davis & Halligan, 2002). Indeed, the prior research usually put on local products production toward the packaging issues on shelf life extension (Alfiana & Latief, 2021; Kontominas et al., 2021), appealing packaging (Butler, 2013; Ying & Anuar, 2019) and the local brand momentum (Hassan & Rahman, 2013). However, the raising question point out from this research is to what extend the prior research lead and consider the potential of packaging and branding as new factor contribute to shape and can empower the human capital among the coastal community?

The empowerment of local knowledge by providing supportive innovation through upgrade local products' packaging, branding, and even the management system, plays an important role to build the customers' trust. Specifically, throughout the ideas of packaging and branding, this conducted study on "ikan bekok" initially confined to traditional wrapping technique which is using coconut leaves. Yet it practically used newspapers due to low cost (Abd. Aziz Itar, 2015), then it gradually changes to plastic packaging. However, packaging without a brand gives less opportunity in expanding market segmentation even marketed domestically. This is due to inadequate nutritious information, shelf-life consuming durations, relevant certificate as to show undoubtful and hygienic preparation and material used.

Therefore, packaging must come with the brand and branding as part of the marketing strategy (Anucha, 2019; Ardhianto & Hardjana, 2016; Hassan & Rahman, 2013; Sebbeh, 2023). In addition, it can be utilized as a revitalization or retain strategy for local products same goes to its local knowledges. Nevertheless, it is seldom to put in the perspective of can human capital being empower by the changes and innovation of branding and packaging. It is a concern that the process of handling, curing, and processing "ikan bekok" to have an additional chain in providing better market value by expanding into the new technological approach of curing the products with the merge of new media platform which refer to online marketing and delivery service.

3 Methodology

The conceptual framework for this study was developed with the implementation of the sequences methods as illustrated in Fig. 1 that involves data collection and information by reviewing prior literature, module delivery, observation, semi—structured interview, proposed innovation on branding and packaging, then displayed into the conceptual framework based on reviewed products marketed. This achievable conceptual framework used the analysis from Miles and Huberman's Flow. It was first introduced by Johnny Saldana in 1984 and it is useful for analyzing qualitative data with recurring relationships through interviews, observations, and other forms of qualitative research (Saldana, 2013). This techniques has been applied in wide range of research as implemented among Minahasa community for identifying the religious reality (Fathimah et al., 2023), maintaining natural resources (Normalina et al., 2020) and exploring knowledge and talent management to deal with the economic recession, then enhance the added value among the involved participants (Andre Suryaningprang et al., 2023; Kurniawan & Aziz, 2023).

The formation of this conceptual framework begins with the determining local knowledge practices among the coastal community in Tumpat, specifically on their activities on curing fish—known as Ikan Bekok and its market. The handling post—harvest resources of a catch simultaneously reviewed thoroughly by the prior and current research. Then, by identifying the participants who are actively practice and producing local food products, 22 participants have been participated in Social Enterprise Program: Ikan Bekok abbreviated as SEP (IB). Only six (6) participants are being interviewed as expert for the research subject (Braun & Clarke, 2019; Guest et al., 2020). The program consists of 14 designated modules that emphasized on three (3) main content which are (1) core module—provided the knowledge transfer related to branding and packaging), (2) Neuro-Linguistic Programming approach for developing the value of human capital and creating teamwork, and (3) entrepreneurship module to provide systematic filing, management, and improvise the business ideas.

Based on Fig. 1, the data were collected and analyzed qualitatively using the Miles and Huberman flow model analysis, which includes three (3) stages (Miles & Huberman, 1994). First stage is Data Reduction—refer to simplifying the overall process of ikan bekok and its environment. Secondly, Data Presentation—it would assemble the first result into the action by proposing brand of "Ikey Bekok Tumpat"

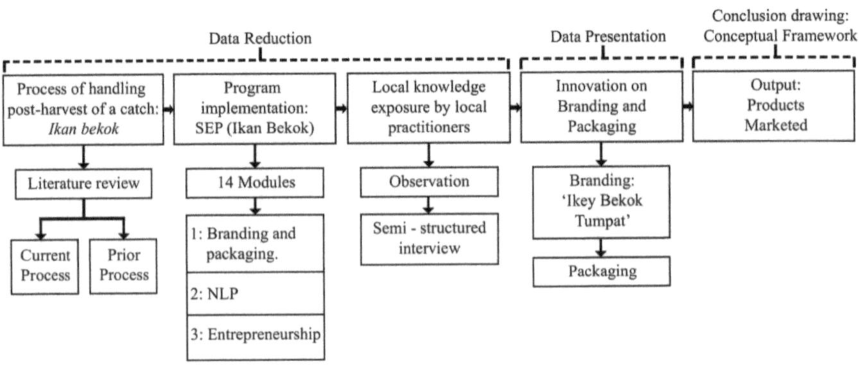

Fig. 1 The sequences methods and analysis involved in collecting data and information for designing the conceptual framework of the study

and its branding component. In addition, proposing plastic pack for ikan bekok as it suitable to conduct with sealing machine or vacuum sealing machine. The third stage is Conclusion Drawing—which tend to illustrate and represent the whole process into the conceptual framework on brand, branding, and packaging potentially become the additional scope in ikan bekok preparation process, at once can elevate and empowering human capital among the coastal community.

4 Findings and Discussions

This study has obtained two (2) final descriptive results including (1) the overall operational chain scope for *ikan bekok* preparation and sales, and (2) the conceptual framework on influence of branding and packaging for local knowledge practices mobilize human capital empowerment.

4.1 Overall Operational Chain Scope for Ikan Bekok Preparation and Sales

This study has found the overall operational chain scope in preparing and market the product of *ikan bekok*. The finding has been illustrated in Fig. 2. Generally, *ikan bekok* process implemented with six (6) phases which preceded with first, preparation of raw material—explained as raw fish collected fresh from sea or river, then neatly eviscerate, split cut the back and clean thoroughly using clean water. Secondly, soaking process needs the cleaned fish to be soaked in brine solution with closed container for overnight or approximately four (4) hours. The third phases is drying process which allowing the brine soaked fish half-dry basking under the sun for 2–4 h on the range known as *Achok* and *Belak*. Then, the fourth phase storage process require ikan bekok to be kept in the big plastic before putting into the freezer as market stock and ready for common marketing which are retail and cuisine.

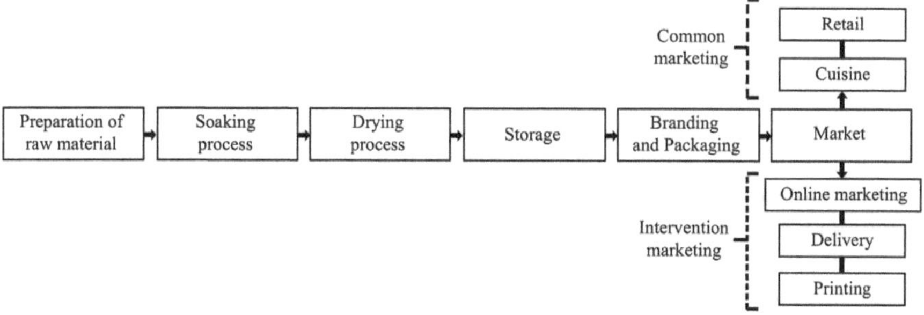

Fig. 2 The overall operational chain scope for local food product of *ikan bekok*

However, the fifth phase involved the branding and packaging—its preferable process on designing the brand with proper nutritious information and attractive graphic. In addition, the packaging priorly used the newspaper and transparent plastic shows less appropriate to prevent excessive water and unpleasant smell, but proposed packaging is suitable to conduct with the sealing machine or vacuum sealing machine. It offers the advantage of freshness, favorable quality of taste, protection, hygienic, food safety, and more organize. Meanwhile, the sixth phase in operational chain scope involved the market. For the common marketing that has been practices by the local practitioners only involved the retail and cuisine service. But intertwin with today's technology and various social media platform, it served the intervention marketing which convenient to use online marketing. It also can provide delivery service opportunities as well as printing for the purpose of attraction as well as giving confidence to customers for buying local perishable products. Therefore, each phase involved in the manufacturing of ikan bekok gradually changed through time, giving employment opportunities to the surrounding population for subsistence.

4.2 *Conceptual Framework of Human Capital Empowerment Through Branding and Packaging Innovation for Local Perishable Product:* Ikan Bekok

In reply to Sustainable Development Goals (SDGs), the global goals covered three (3) pillars of social (56%), economic (20%), and environment (24%) with implementing specific target. The new agent of changes would be beneficial to the holistic perspective of business, government, and society. There is strong emphasize on understanding toward the various approaches, concepts, and mechanisms in accomplishing the SDGs with effective comprehend contribution. Based on Fig. 3, the elaboration on the conceptual framework on how branding and packaging related with the implementation of social enterprise, empowering human capital and at once sustaining the local knowledge as it comply with the Sustainable Development Goals (SDGs) referring to quality education (SDGs 4) and decent work and economic growth (SDGs 8).

SDGs 4: Quality education—refer to quality education for all which preferable the insight of merging with the formal education and informal education. The emergence of knowledge transmission from institution and local knowledge from the community mutually interrelated (and support each other) to sustain local gastronomic products. Meanwhile, SDGs 8: Decent work and economic growth—involved three (3) phases which are Phase 1: Manufacturing process is the process of manufacturing productivity. It provides the operational chain scope which allow the single mother and residents to participate in the job lines. Phase 2 is branding and packaging. Based on this study, the term of "branding" may adapt as it generally refers inclusively related to brand, branding, and packaging process of ikan bekok

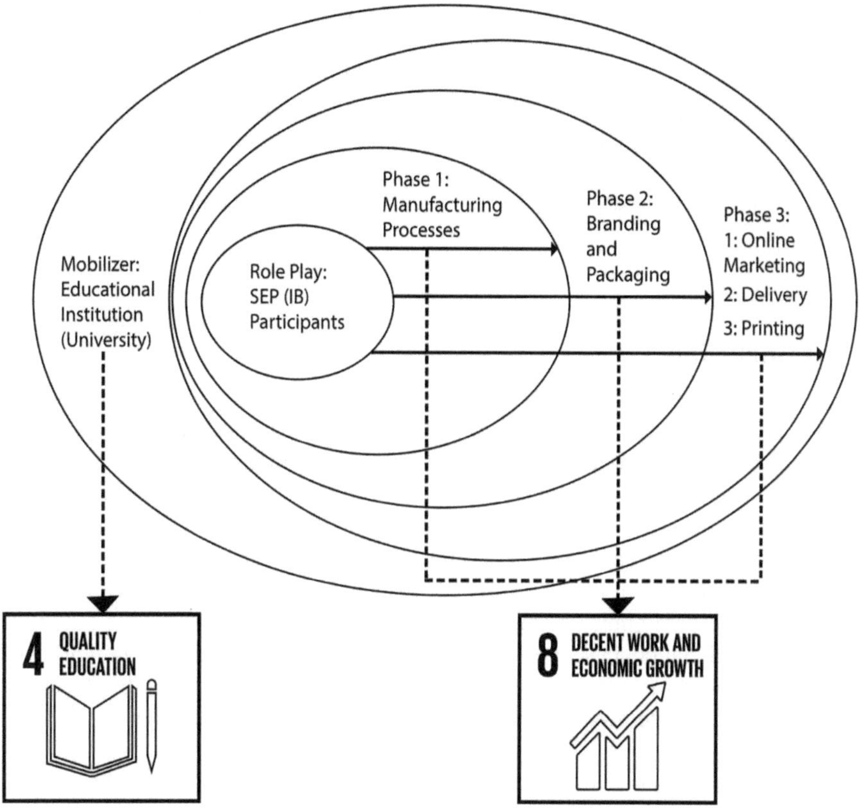

Fig. 3 The illustration on expanding job creation in coastal community in Tumpat involved the educational institutions as the activator to mobilize and expand the sub jobs opportunity

products properties. This technological innovation proposed branding and packaging for local perishable products by following the standard operational procedure. Lastly, Phase 3: Online Marketing, Delivery, and Printing included in job creations. Due to uncertain condition in marketing local perishable products of ikan bekok and the rises of interest among youth using the digital and online platform, it is an opportunity for online marketer job to promote and sell the local products. While it is a need for delivery service to meet the current demand and printing service for designing and designated the attractive packaging.

Therefore, the innovation on branding and packaging toward the local product of ikan bekok indicate three (3) key values which are (1) visual locality representative, (2) optimizing gastronomic values, and (3) expanding domestic economic growth. Firstly, visual locality representative by giving the branding of "Ikey Bekok Tumpat" represent the local products of ikan bekok produce by the coastal community in Tumpat, Kelantan. Secondly, optimizing gastronomic values—by proposing packaging with the brand "Ikey Bekok Tumpat" optimizing the gastronomic delight for wide market demand (local and aim for international market) with guaranteed from local productions. Besides, the propose packaging provides the preferences for better packaging with hygienic, freshness control, and keep nutritious for perishable products. Thirdly, expanding domestic economic growth and empowering human capital. The implementation of SEP (IB) gives an exposure in sustaining the social enterprise products which are (1) by serving the food resources for people, and (2) expanding the domestic growth and subsistence living cost for surrounding residents through local wisdom practices entwined with the influence of branding and packaging.

5 Conclusion

Post-harvest handling of fisheries products is become part of cumulative skills and practices of local wisdom by the coastal community in East Coast of peninsular Malaysia. The intertwined transmission knowledge between formal and informal education has trigger the human capital empowerment by creating the job expansion along as their social enterprise. Therefore, this study conclude that branding and packaging can be one of the factors that contribute to empower the human capital by combining the community strength in producing local organoleptic food and the current operational chain scopes. Furthermore, this is the strategy that comply with SDGs which lead to economic development and universal well-being for achieving stability in the perspective economic, society, and environmental pillars.

Acknowledgement This research acknowledges Malaysia Ministry of Finance for providing financial support under UMK – MoF Social Enterprise Project, Code: (R/MOF/A0200/00373A/003/2020/ 00733) collaborates with Universiti Malaysia Kelantan (UMK).

References

Abd. Aziz Itar. (2015). Bahaya Kertas Surat Khabar sebagai Pembungkus dan Penutup Makanan: Beri Kesedaran Pengguna. *Utusan Malaysia, 18*.

Alfiana, S. H., & Latief, A. D. R. (2021). The potential of active packaging for tuna. *International Conference of Animal Science and Technology, 788*, 012227. https://doi.org/10.1088/1755-1315/788/1/012227

Anucha, V. C. (2019). *African Journal of Innovations in Marketing Research and Management, 8*(2), 9–23.

Ardhianto, P., & Hardjana, A. C. (2016). Smes traditional packaging technology as a part of building brand soul of product. *International Journal of Computational Engineering Research (IJCER), 6*(07), 10–15.

Aulia, F., Badaruddin, R., Harahap, H., & Utomo, B. (2020). Local knowledge of North Sumatera Coastal Communities on sustainable Mangrove identification and types. *Journal of Southwest Jiaotong University, 55*(3).

Berkes, F., & Davidson-Hunt, I. J. (2007). Communities and social enterprise in the age of globalization. *Journal of Enterprising Communities: People and Places in the Global Economy, 1*(3), 209–221. https://doi.org/10.1108/17506200710779521

Boateng, D. A. (2023). Women Traders at the Elmina Fishing Harbor: Harbingers of Ghana's Social Development. In *Women's Contributions to Development in West Africa* (pp. 57–72). Springer Nature Singapore.

Braun, V., & Clarke, V. (2019). To saturate or not to saturate? Questioning data saturation as a useful concept for thematic analysis and sample - size rationales. *Qualitative Research in Sports, Exercise and Health, 13*(21), 201–216. https://doi.org/10.1080/2159676X.2019.1704846

Butler, P. (2013). Smart and interactive packaging developments for enhanced communication at the packaging/user interface. In *Trends in Packaging of Food, Beverages and Other Fast-Moving Consumer Goods (FMCG)* (pp. 261–287). https://doi.org/10.1533/9780857098979.261

Cheah Sau Seng. (2018). *The determinants of social enterprises performance in Malaysia and Singapore*. Universiti Sains Malaysia.

Choudhary, M.-U.-I., Emdad Haque, C., Nishat, A., & Byrne, S. (2021). Social learning for building community resilience to cyclones: Role of indigenous and local knowledge, power, and institutions in Coastal Bangladesh. *Ecology and Society, 26*(1), 5.

Davis, S., & Halligan, C. (2002). Extending your brand by optimizing your customer relationship. *Journal of Consumer Marketing, 19*(1), 7–11. https://doi.org/10.1108/07363760210414916

Douglas, H., Eti-Tofinga, B., & Singh, G. (2018). Contextualising social enterprise in Fiji. *Social Enterprise Journal, 14*(2), 208–224. https://doi.org/10.1108/SEJ-05-2017-0032

Ejere, E. S. I. (2011). Human capital formation as catalyst for national development: Nigeria in perspective. *International Business and Management, 2*(2), 98–104. www.cscanada.org

Fathimah, S., Kerebungu, F., Sidik, S., & Suarjana, I. W. G. (2023). Minahasa Community Local Wisdom; Study of Social Construction of Inter-religious Harmony. *Unima International Conference on Social Sciences and Humanities, 1*, 1731–1738. https://doi.org/10.2991/978-2-494069-35-0

Frangoudes, K., & Keromnes, E. (2008). Women in Artisanal Fisheries in Brittany, France. *Development, 51*, 265–270. https://doi.org/10.1057/dev.2008.20

Guest, G., Namey, E., & Chen, M. (2020). A simple method to assess and report thematic saturation in qualitative research. *PLoS One, 15*(5), 1–17. https://doi.org/10.1371/journal.pone.0232076

Haji Megat Abu Bakar, M. M. F. (2022). Satay fish and other traditional fish products development in Malaysia. In *Development of fish products* (pp. 87–90). Marine Fisheries Research Department.

Hassan, H., & Rahman, M. S. (2013). Local brand momentum through hypermarket channel. *Management, 3*(2), 89–92. https://doi.org/10.5923/j.mm.20130302.04

Hua, G. B. (1989). Present status of fish processing in Malaysia. In *Advances in Fishery Post - Harvest Technology* (pp. 98–102). Marine Fisheries Research Department.

Huffman, D., & Quigley, J. M. (2002). The role of the university in attracting high tech entrepreneurship: A Silicon Valley tale (No. W01-006). *The Annals of Regional Science, 36*, 403–419.

Kim, D., & Sturman, M. (2012). HR branding: How human resources can learn from product and service branding to improve attraction, selection, and retention.

Kontominas, M. G., Badeka, A. V., Kosma, I. S., & Nathanailides, C. I. (2021). Recent developments in seafood packaging technologies. *Foods, 10*(5), 940. https://doi.org/10.3390/foods10050940

Kurniawan, H., & Aziz, A. (2023). Ab-Chicken Sharia Partnership Business Strategy: In the Face of a Global Economic Recession. *Jurnal Ekonomi Syariah Indonesia, 13*(1), 24–35.

Maguirre, M. V., Portales, L., & Bellido, I. V. (2017). Indigenous social enterprise as drivers of sustainable development: Insight from Mexico and Peru. *Critical Sociology, 44*(2), 323–340. https://doi.org/10.1177/0896920516688757

Marimuthu, M., Arokiasamy, L., & Ismail, M. (2009). Human capital development and its impact on firm performance: Evidence from developmental economics. *Journal of International Social Research, 2*(8), 265–272. http://scholar.google.com/scholar?hl=en&btnG=Search&q=intitle:No+Title#0

Marion, A. (2021). What is branding? *The Branding Journal*. https://www.thebrandingjournal.com/2015/10/what-is-branding-definition/

Marsden, T., & Smith, E. (2005). Ecological entrepreneurship: Sustainable development in local communities through quality food production and local branding. *Geoforum, 36*, 440–451. https://doi.org/10.1016/j.geoforum.2004.07.008

Mei, W., Yee, T., & Rahman, S. (2019). Empowerment for economic and human capital development through education. In *ASEAN Post-50*. Springer Singapore. https://doi.org/10.1007/978-981-13-8043-3

Miles, M. B., & Huberman, A. M. (1994). *Qualitative data analysis* (2nd ed.). SAGE Publications.

Normalina, H., Hatta, M., & Hamdani, H. (2020). Green Leadership "the forest guard" from the upstream. *EurAsian Journal of Biosciences, 14*, 5835–5842. https://repo-dosen.ulm.ac.id/bitstream/handle/123456789/26840/GREEN_LEADERSHIP_"THE_FOREST_GUARD"_FROM_THE_UPSTREAM.pdf?sequence=1&isAllowed=y

Nuridayati, M. R., & Hasan, K. (2021). Empowerment communication as a new perspective in the empowerment of coastal communities of Pase Raya. In *Proceeding of the international conference on socail science, political science, and humanities* (pp. 24–31).

Peloquin, C. (2009). Local knowledge, subsistence harvests, and social-ecological complexity in James Bay. *Human Ecology, 37*(5), 533–545. https://doi.org/10.1007/s10745-009-9255-0

Rahmawati, A., & Apip, A. (2022). Partnership model for the implementation of corporate social responsibility state own company in coastal community empowerment programs in Cilegon City. In *Universitas lampung international conference on social sciences* (pp. 538–543).

Rajab, M., Nawawi, J., & Rusli, A. M. (2022). The Role of the Marine and Fisheries Office in Empowering Fishermen in Majene Regency. In *Universitas Lampung International Conference on Social Sciences* (pp. 169–174).

Rhoumah, A. M. O. (2016). Determinants of factors that affect poverty among coastal fishermen community in Malaysia. *Journal of Economics and Finance, 7*(3), 9–13. https://doi.org/10.9790/5933-0703020912

Rossiter, J. R. (2014). "Branding" explained: Defining and measuring brand awareness and brand attitude. *Journal of Brand Management, 21*(7–8), 533–540. https://doi.org/10.1057/bm.2014.33

Ruddle, K. (1994). Local knowledge in the folk management of fisheries and coastal marine environments. In *Folk management in the world's fisheries* (pp. 161–206).

Saad, M. (2020). Key elements of nation branding: The importance of the development of local human capital in the UAE. In V. Pereira, M. Neal, Y. Temouri, & W. Qureshi (Eds.), *Human capital in the middle east* (pp. 227–251). Palgrave Macmillan. https://doi.org/10.1007/978-3-030-42211-0_9

Saldana, J. (2013). *The coding manual for qualitative researchers* (2nd ed.). SAGE.

Sebbeh, B. J. (2023). The effects of packaging design on the marketability of 'made in Ghana' products. *International Journal of Innovative Research and Development, 11*(8), 66–77. https://doi.org/10.24940/ijird/2022/v11/i8/aug22031

Sembok, T. M. Z. T., & Ismail, W. A. A. Z. W. (2021). Formal education, work experience and empowerment of the fishing community in Terengganu, Malaysia. *Journal of Nusantara Studies, 7*(1), 2013–2015. https://doi.org/10.24200/jonus.vol7iss1pp366

Šlaus, I., & Jacobs, G. (2011). Human capital and sustainability. *Sustainability, 3*(1), 97–154. https://doi.org/10.3390/su3010097

Smith, A. (2007). An inquiry into the nature and causes of the wealth of nations. In *MetaLibri*.

Steiner, A., Farmer, J., & Bosworth, G. (2019). Rural social enterprise - Evidence to date and a research agenda. *Journal of Rural Studies, 70*, 139–143. https://doi.org/10.1016/j.jrurstud.2019.08.008

Suryaningprang, A., Saputro, A. H., Herlina, L., & Supiyadi, D. (2023). Knowledge-based management and talent management as efforts to increase the economic value-added of universities. *European Journal of Military Studies, 13*(2023), 2656–2666.

Vayda, A. P., Walters, B. B., & Setyawati, I. (2019). Doing and knowing: Questions about studies of local knowledge. In *Investigating Local Knowledge* (pp. 35–58). Routledge.

Ying, C. S., & Anuar, T. F. T. (2019). Packaging innovation as a commodification excellence factor for small and medium enterprises (SMEs) case study: Frozen Keropok Lekor packaging in Kelantan. *International Journal of Entrepreneurship and Management Practices, 2*(5), 01–15.

Zainon, S., Ahmad, S. A., Atan, R., Wah, Y. B., Bakar, Z. A., & Sarman, S. R. (2014). Legitimacy and sustainability of social enterprise: Governance and accountability. *Procedia - Social and Behavioral Sciences, 145*, 152–157. https://doi.org/10.1016/j.sbspro.2014.06.022

Zal, W. A. A. (2008). The ownership and use of new human capital among Orang Kuala in Malaysia and its relation with economic empowerment. *International Proceedings of Economics Development and Research (IPEDR)*, 162–166.

Zamroni, A., Amano, M., & Yamao, M. (2017). A brief comparison of fisheries livelihoods strategy: Countries' experiences of Japan and Indonesia. *International Journal of Marine Science, 7*(38), 371–379. https://doi.org/10.5376/ijms.2017.07.0038

Tengku Fauzan Tengku Anuar is a Senior Lecturer in the Department of Visual Communication, Faculty of Creative Technology and Heritage, Universiti Malaysia Kelantan. She has been teaching for twelve years at the university. She teaches Visual Communication Design to bachelor's degree students and supervises master's and PhD candidates during her university service until the present. Her educational background in Visual Communication and New Media has given her a broad base from which to approach many topics. Her research skills may be confirmed independently, and getting more science, she joins the University Centre of Excellence (COE), Institute for Poverty Research and Management (InsPeK). She especially enjoys serving the community through research and capacity building development programs. She believes that the meaning of life is giving back to others in need.

Nurul Syahida Mat Hussin is a research assistant at the Institute of Research and Poverty Management, involved with local community development since 2019, and currently pursuing PhD program in Creative Technology Degree at Universiti Malaysia Kelantan, Malaysia.

Ren Chen Yang, graduated from Shanxi Datong University, bachelor degree, engaged in design training, education and training related work, with professional knowledge and skills, excellent creative ability, good at analyzing customer psychology, good at communication. He is currently pursuing a master's degree in visual communication at the University of Kelantan, Malaysia.

An Empirical Study on Green Entrepreneurship and Sustainable Entrepreneurship Education Development

Siti Afiqah Zainuddin, Tahirah Abdullah, Amira Jamil, Nur Izzati Mohamad Anuar, Liyana Ahmad Afip, Siti Rohana Mohamad, Nik Malini Nik Mahdi, Muhammad Zahid Zainuddin, and Siti Hanis Zainuddin

Abstract Consumers today are more concerned about the environment and are moving towards a greener market, resulting in a rise in demand for environmentally responsible products. In this regard, the current study aims to examine the impact of green entrepreneurship on the sustainable development of entrepreneurship education among undergraduates at Universiti Malaysia Kelantan (UMK). This study reviews existing literature on green entrepreneurship and sustainable entrepreneurship education development before constructing a research framework. A questionnaire is then used to collect primary data. Three hundred forty-six data were collected from Universiti Malaysia Kelantan students via an online survey using a

S. A. Zainuddin
Faculty of Entrepreneurship and Business, UMK, Kota Bharu, Kelantan, Malaysia

Department Accounting, Economics and Finance, School of Business Administration, European University Cyprus, Engomi, Cyprus
e-mail: sitiafiqah@umk.edu.my

T. Abdullah (✉) · A. Jamil · N. I. M. Anuar · S. R. Mohamad · N. M. N. Mahdi
Faculty of Entrepreneurship and Business, UMK, Kota Bharu, Kelantan, Malaysia
e-mail: tahirah@umk.edu.my; amira@umk.edu.my; nurizzati.ma@umk.edu.my; rohana@umk.edu.my; nmalini@umk.edu.my

L. A. Afip
Faculty of Language Studies and Human Development, UMK, Kota Bharu, Kelantan, Malaysia
e-mail: liyana.a@umk.edu.my

M. Z. Zainuddin
Ministry of Entrepreneurship and Cooperatives Development (KUSKOP), Putrajaya, Malaysia
e-mail: zahid.zainuddin@kuskop.gov.my

S. H. Zainuddin
Petroliam Nasional Berhad (PETRONAS), Kuala Lumpur, Malaysia
e-mail: sitihanis.zainuddin@petronas.com

N. Hayat, N. C. Nawi (eds.), *Contemporary Issues in Entrepreneurship and Innovative Technology*, Contributions to Environmental Sciences & Innovative Business Technology, https://doi.org/10.1007/978-3-031-50927-8_5

convenience sample. To analyse the data, the Pearson correlation coefficient and a reliability test were conducted using Statistical Package for the Social Sciences (SPSS). The findings indicate that the independent variables, namely, the professional development of entrepreneurship education, the integration of sustainability into the academic curriculum, an increase in financial support, enhancement towards establishing and straightening green culture, and the expansion of green buildings and facilities, are significant to the UMK study. Lastly, the effect of the green market on sustainable development and green entrepreneurship has not been thoroughly studied. This study aims to highlight the existing literature on green sustainable development education for entrepreneurs.

Keywords Entrepreneurship education · Green entrepreneurship · Quantitative analysis · Entrepreneurship student

1 Introduction

The green education model broadens the scope of services to include the types of education that organisations provide (the process of producing services), how those services are communicated to clients, and how service providers can benefit from providing those services (value realisation). Green entrepreneurial education is based on green education, which is believed to represent the formation and development of competencies for long-term development, such as the knowledge, skills, and behaviours required for sustainable living and behaviour, according to current research (Kotla & Bosman, 2023). Entrepreneurship education plays a vital role in promoting economic development, and graduates of entrepreneurship programmes are vital contributors to the dynamism and adaptability of corporate environments (Boldureanu et al., 2020). After graduation, for instance, undergraduates are capable of innovating concepts based on their knowledge to create unique business opportunities, innovations in the organisation of their employment, and more.

This research intends to examine the impact of green entrepreneurship on the sustainable development of entrepreneurship education among undergraduates at Universiti Malaysia Kelantan. The concept of sustainable development in entrepreneurship education has emerged, emphasizing the concurrent growth of general and professional skills, encompassing both knowledge acquisition and skill enhancement. Shulla et al. (2020) describe that many institutions concentrate on Sustainable Development Goal (SDG) 4: Quality Education; this is due to the significance of SDGs in emerging entrepreneurship education, where it is essential for bringing the curriculum to life by empowering students.

This study aims to examine the relationship between (1) the professional development of green entrepreneurship education and the sustainable development of green entrepreneurship education, (2) the integration of sustainability into academic curricula and the sustainable development of green entrepreneurship education, (3) the increase in financial support for green entrepreneurship and the sustainable development of green entrepreneurship education, and (4) the boost in financial support for green entrepreneurship and the sustainable development of green entrepreneurship education.

2 Literature Review

Economic and human development growth are correlated to sustainable development. Consequently, the primary objective of this literature review is to investigate the underlying theory underlying the design study of emotional intelligence and job performance. This study employs the Owston theory and draws from the Owston sustainability of innovation education model for its purposes (Owston, 2004). The model is adopted and modified to fit the current context of the study.

2.1 Professional Development of Green Entrepreneurship

Green entrepreneurship has gained significant importance over the years due to the increasing environmental challenges and the need for sustainable development. The professional development of green entrepreneurship is essential for ensuring that entrepreneurs acquire the necessary knowledge, skills, and attitudes to establish and run successful green businesses.

Investing in teacher professional development in green entrepreneurship education is crucial to improve student learning. According to the concept, when teachers realise that their students support innovation and that it enhances their learning, they are more likely to devote time and effort to ensuring its success. As teachers invest more in innovation, they learn more about the pedagogical methods they employ, such as project-based learning and sustainable development.

In another aspect, Zhang et al. (2019) examined the role of government policies in promoting the professional development of green entrepreneurship. The study found that government policies play a significant role in promoting green entrepreneurship by providing incentives and support to entrepreneurs pursuing professional development in green entrepreneurs. Supported by Chen et al. (2018), who investigated the impact of training on the professional development of green entrepreneurs, the study found that training significantly improved green entrepreneurs' knowledge, skills, and attitudes, leading to successful green businesses. Li et al. (2020) investigate the influence of social capital on the professional development of green entrepreneurship. The study found that social capital positively influenced the professional development of green entrepreneurship by providing access to resources, knowledge, and support.

In addition, Guo et al. (2021) examined the importance of mentoring in the professional development of green entrepreneurs. The study found that mentoring positively influenced the professional development of green entrepreneurs by providing guidance, support, and networking opportunities. The mentor could be obtained from the local and global market to increase the quality and scope of mentoring. Wang et al. (2020) study the impact of internationalisation on the professional development of green entrepreneurship. The study found that internationalisation positively influenced the professional development of green entrepreneurship by providing access to new markets, resources, and knowledge. The previous studies highlight the importance of green entrepreneurship's professional development in promoting sustainable development. Government policies, training, social capital, mentoring, and internationalisation were identified as key factors influencing the professional development of green entrepreneurship. Therefore, a hypothesis is developed:

H1: There is a significant relationship between the professional development of green entrepreneurship education and sustainable entrepreneurship education development.

2.2 *Integration of Sustainability Agenda in Entrepreneurship Academic Curriculum*

The incorporation of sustainability into the academic curriculum can connect innovation and students. In addition to the new curriculum reform for basic education, all provinces within the institution have examined "new sustainable education" for undergraduates, with "innovation" as the primary objective. Wu et al. (2020) analyses and evaluates the impact of curricular innovation on the growth of students' learning capacity. The study found that curriculum innovation should be defined in the current curriculum instruction from the perspective of cultivating students' creativity and learning abilities to support the growth and improvement of students' learning capacity.

Incorporating sustainability into academic entrepreneurship curricula has become increasingly important due to the growing need for sustainable development. Adegbite et al. (2020) highlight the importance of integrating sustainability into entrepreneurship education. The study found that integrating sustainability into entrepreneurship education could lead to the development of sustainable businesses and a sustainable future. However, there are challenges in this attempt. Al-Tabbaa and Ankrah (2019) opine that there are challenges to integrating sustainability into entrepreneurship education. The study found that the lack of awareness, resources, and expertise was one of the key challenges faced in integrating sustainability into entrepreneurship education.

Despite the challenges, due to the high impact of sustainability education on entrepreneurial intention, it has to be paid more attention. Al-Harthi and Alghamdi (2020) found that sustainability education positively influenced entrepreneurial intention and the likelihood of starting sustainable businesses. Integrating sustainability into entrepreneurship education is a way of realising the objective. Fernandes and Castelo Branco (2021) examine the role of experiential learning in integrating sustainability into entrepreneurship education. The study found that experiential learning, such as internships and case studies, provided students with the practical knowledge and skills needed to address sustainability challenges in their future businesses, supported by the study of Guzmán and Reyes-Rodríguez (2018), who found that sustainability-based courses positively impacted students' attitudes towards sustainability and increased their understanding of the role of entrepreneurship in sustainable development. Therefore, a hypothesis is developed:

H2: There is a significant relationship between integrating sustainability agenda in entrepreneurship academic curriculum and the sustainable development of green entrepreneurship education.

2.3 Increase in Financial Assistance

The increase in financial support can be correlated to the external support described by Potluka & Svecova (2019). The correlation between unequal institution funding and unequal educational outcomes, such as lower achievement and graduation rates, may result in fewer undergraduate admission opportunities for severely underfunded institutions. This exceeds the basic academic needs of students, such as access to modern technology, sanitary and secure school facilities, and classroom materials. Education also includes activities outside the classroom, such as enrichment programmes in the arts; science, technology, engineering, and mathematics (STEM) sectors; and other disciplines. By increasing financial support, university fundraising increases the number of students who can benefit from these opportunities and promotes a more comprehensive education. While there are many fantastic fundraisers, those involving student participation, such as enjoyable races, have the added benefit of teaching participants valuable life skills and habits. Therefore, a hypothesis is developed:

H3: There is a significant relationship between the increase in financial support and the sustainable development of green entrepreneurship education.

2.4 Improvement Towards Harmonising Green Culture

Gunawan and Widodo (2020) highlighted the role of green entrepreneurship culture in promoting sustainable development. The study found that a strong green entrepreneurship culture could lead to the development of sustainable businesses and a sustainable future. There are several initiatives already in place to promote green culture. Singh and Jain (2020) examined the role of incubation programmes in enhancing green entrepreneurship culture. The study found that incubation programmes provided the support and resources needed for entrepreneurs to develop sustainable businesses and promoted the growth of green entrepreneurship culture.

The government promotes the culture at the higher administration level through policy development. Aslani and Karimi (2019) highlighted the importance of government policies in promoting green entrepreneurship culture. The study found that supportive policies, such as tax incentives and funding programmes, could encourage entrepreneurs to develop sustainable businesses and contribute to sustainable development. In the wider community, Shahab and Rezazadeh (2020) study the role of social networks in promoting green entrepreneurship culture. The study found that social networks provided opportunities for entrepreneurs to exchange ideas, share experiences, and develop partnerships, contributing to the green entrepreneurship culture's growth. Therefore, a hypothesis is developed:

H4: There is a significant relationship between the enhancement of green entrepreneurship culture and the sustainable development of green entrepreneurship education.

2.5 Expansion of Green Facilities and Infrastructure

Green infrastructure and facilities are crucial in promoting sustainable development and green entrepreneurship education. Anwar and Shaari (2021) examine the role of green infrastructure in promoting sustainable development. The study found that green infrastructure, such as parks and green roofs, could help reduce urban heat islands, improve air quality, and promote sustainable living. Regarding education, green infrastructure could provide opportunities for students to learn about sustainable practices and develop sustainable business ideas (Lee & Chae, 2018). Elgizawy (2019) examined the role of sustainable facilities in promoting green entrepreneurship education. The study found that sustainable facilities, such as green buildings and renewable energy systems, could provide a learning environment that promotes sustainability and prepares students for sustainable entrepreneurship.

Providing green facilities and infrastructure to promote the sustainable development of entrepreneurship education is not a simple task, and it needs public and private cooperation. Krometis and de Souza (2019) highlighted the importance of public-private partnerships in expanding green infrastructure and facilities. The study found that public-private partnerships could provide the funding and resources needed to develop green infrastructure and facilities and promote sustainable development. Another important aspect of public body intervention is developing and introducing supportive policies, such as tax incentives and regulations, which could encourage investment in green infrastructure and facilities and promote sustainable development (Fu & Yu, 2020). Therefore, a hypothesis is developed:

H5: There is a significant correlation between the expansion of green infrastructure and facilities and the sustainable growth of green entrepreneurship education.

2.6 Research Framework

A research framework was developed to depict a clear picture of the relationship between independent and dependent variables. Relevant variables derived from the literature are incorporated into this model, and each variable illustrates the relationship between the remaining variables. The following conceptual model has been developed based on previous research reported in the literature review. The conceptual model investigates the relationship between green entrepreneurship education and sustainable development. The dependent variable is the sustainable development of entrepreneurship education, while independent variables are the professional development of entrepreneurship education, the integration of sustainability into the academic curriculum, an increase in financial support, improvement towards establishing and straightening green culture, and the expansion of green buildings and facilities. The research framework for this investigation is displayed in Fig. 1.

3 Research Methodology

3.1 Research Design

The study of design is generally a study plan that outlines the processes that other researchers should take seriously to achieve their research objectives or test the hypotheses they have suggested for their research. This study used quantitative

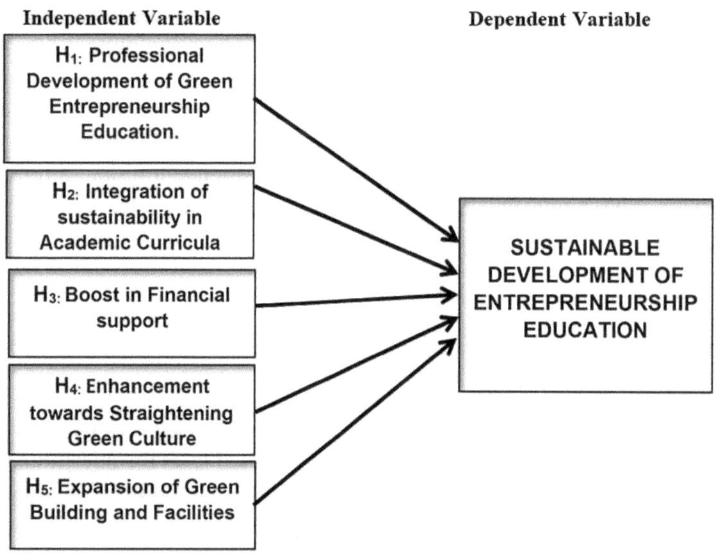

Fig. 1 Research framework

research, and data were collected using an online survey. According to Hair, Black, Babin, and Anderson (2006), an online survey can help decrease errors during the process. This research design integrates data collection and analysis phases within a quantitative framework, chosen for its suitability in measuring and understanding observable behavior. Descriptive quantitative research will be the focus.

3.2 Collection of Data

Google Forms was used to create the online survey. The link to the online questionnaire was then disseminated through social media platforms, such as WhatsApp, and the researcher's personal contacts. There were 346 participants in this study. The questionnaires include three sections, labelled A, B, and C. Section A consists of demographic questions such as gender, age, academic programme, and year of study. Sections B and C measure independent and dependent variables using a five-point Likert scale ranging from 1 (strongly disagree) to 5 (strongly concur).

3.3 Sampling Method

In research methodology, the term "population" designates the complete set of individuals, objects, or events that are of interest to the researcher and from which data are collected for analysis and generalization. In statistics, a population refers to the entire collection of statistical samples (Sekaran & Bougies, 2016). Probability sampling would be chosen because it serves to clarify the relationship between the sample and the population. Therefore, straightforward random sampling is a probability sampling technique that involves selecting a student respondent to answer questionnaires. Everyone in the target population is equally likely to be included in the sample when probability sampling techniques are utilised. Consequently, the student population on this campus was chosen as the respondent frame sample for this investigation. The sample will consist of 346 undergraduate students from Universiti Malaysia Kelantan.

3.4 Method of Analysis

The researcher will use Statistical Package for the Social Sciences (SPSS) to assess the findings of the study by analysing the data collected from the questionnaire after they have been published online as a Google Form. In addition, data analysis is crucial for elucidating the result in order to answer the research query. The Statistical Package for the Social Sciences (SPSS) software is used to analyse the data collected from the distributed questionnaires in this study. In the analysis phase of this study, descriptive statistics, the Spearman correlation coefficient, the normality test, and multiple linear regression will all be employed.

4 Analysis, Results, and Discussion

4.1 Descriptive Analysis

According to Table 1, 346 respondents answered the surveys using the Google Form platform, a summary of all respondents' demographic profiles. Four questions based on each participant's profile, including gender, age, academic programme, and year of study, made up section A of the data.

The result from Table 1 shows that the most respondent by gender is from the female group, which consists of 177 (50.4%) respondents. For the age range, most respondents, 186 (53.0%), are aged 21–25. After that, the least respondent is from the age group above 25 years, consisting of 68 (19.4%) respondents. Next, for the year of study, most respondents are from year 3, which includes 124 respondents

Table 1 Summary of the respondents' demographic profile

Items	Frequency	Percentage (%)
Gender		
Male	174	49.6
Female	177	50.4
Total	351	100
Age		
18–20 years	97	27.6
21–25 years	186	53.0
Above 25 years	68	19.4
Total	351	100
Year of study		
Year 1	40	11.4
Year 2	120	34.2
Year 3	124	35.3
Year 4	67	19.1
Total	351	100
Course		
SAE	40	11.4
SAL	66	18.8
SAB	75	21.4
SAK	81	23.1
SAR	65	18.5
SAA	24	6.8
Total	351	100

Table 2 Reliability analysis correlation

Cronbach's alpha range	Level of reliability
a > 0.9	Excellent
a > 0.8	Good
a > 0.7	Acceptable
a > 0.6	Questionable
a > 0.5	Poor
a > 0.4	Unacceptable

(35.3%), and the least is from year 1, representing 40 respondents (11.4%). Lastly, for the course, most are from the SAK course, with 81 respondents (23.1%), and the least is from SAA, with 24 respondents (6.8%).

4.2 Reliability Test

Table 2 shows the results of the reliability of Cronbach's alpha.

Table 3 revealed that all variables were greater than the value of 0.7. This table indicates that the internal consistency of the dependent variable, sustainable development of entrepreneurship education, is moderate (four items = 0.765). And for the independent variables, the professional development of green entrepreneurship education has the highest value (six items = 0.838) among the other independent variables, followed by an increase in financial assistance, which has a value = 0.813 (five items). The Cronbach alpha value for incorporating sustainability into academic curriculum is 0.812 (six items), while the value for improving green culture is 0.810 (five items). Last, the expansion of green buildings and facilities has the lowest value (four elements = 0.788) among the other variables but is still acceptable. Consequently, the Cronbach alpha values derived from these variables are all acceptable and reliable, and they can be used to disseminate the questionnaire to the respondents.

Table 3 Result of reliability Cronbach's alpha

Variables	Cronbach's alpha	Number of items	Strength
Sustainable development of entrepreneurship education (DV)	0.765	4	Moderate
Professional development of green entrepreneurship education (IV)	0.838	6	Good
Integration of sustainability In the academic curriculum (IV)	0.812	6	Good
Boost in financial support (IV)	0.813	5	Good
Enhancement towards straightening green culture (IV)	0.810	5	Good
Expansion of green Buildings and facilities (IV)	0.788	4	Moderate

4.3 Pearson's Correlation Coefficient

Based on Table 4, it is shown the result of Pearson's correlation coefficient. Pearson's correlation coefficient was used to test the association between the hypothesis for independent variables. These are professional development of entrepreneurship education, integration of sustainability in academic curriculum, boost in financial support, enhancement towards establishing and straightening green culture, and expansion of green buildings and facilities, and for the dependent variable is the sustainable development of entrepreneurship education among undergraduate students in Universiti Malaysia Kelantan (UMK).

The findings show that professional development in entrepreneurship education, integrating sustainability into the curriculum, increased financial support, and efforts to establish and strengthen a green culture all have a significant positive impact on the sustainable development of green entrepreneurship education. Additionally, a moderate positive relationship was identified between these factors and the expansion of green buildings and facilities. The relationship between the independent and dependent variables will be stronger when the alpha Cronbach values are between 0.7 and 1; thus, hypothesis 1, hypothesis 2, hypothesis 3, hypothesis 4, and hypothesis 5 are supported.

The objective of this study is to examine the relationship between the sustainable development of entrepreneurship education based on the professional development of green entrepreneurship education, the integration of sustainability in academic curricula, boost in financial support, enhancement towards establishing and straightening green culture, and the expansion of the green building and facilities towards undergraduates of Universiti Malaysia Kelantan (UMK). Based on the findings, the researcher agreed to obtain all the information required for this research, and the questionnaire was distributed online. All hypotheses are supported based on the result presented in Table 5.

Table 4 Pearson correlation

	Std. deviation
Sustainable development of entrepreneurship education	0.629
Professional development of entrepreneurship education	0.577
Integration of sustainability in academic curriculum	0.641
Boost in financial support	0.622
Enhancement towards establishing and straightening green culture	0.652
Expansion of green building and facilities	0.652

Table 5 Results of Pearson correlation coefficient and hypotheses testing

Hypothesis	Significance value	Findings
H1: There is a significant relationship between the professional development of green entrepreneurship education and sustainable entrepreneurship education development.	0.838 Good relationship	H1: Supported
H2: There is a significant relationship between the integration of sustainability agenda in entrepreneurship academic curriculum and the sustainable development of green entrepreneurship education.	0.812 Good relationship	H2: Supported
H3: There is a significant relationship between the increase in financial support and the sustainable development of green entrepreneurship education.	0.813 Good relationship	H3: Supported
H4: There is a significant relationship between the enhancement of green entrepreneurship culture and the sustainable development of green entrepreneurship education.	0.810 Good relationship	H2: Supported
H5: There is a significant correlation between the expansion of green infrastructure and facilities and the sustainable growth of green entrepreneurship education.	0.788 Moderate Relationship	H5: Supported

5 Discussions and Conclusion

The primary goal of this study is to investigate the relationship between the effect of green entrepreneurship on the sustainable development of entrepreneurship education and the five independent variables tested, namely, professional development of green entrepreneurship education, integration of sustainability in academic curricula, increase in financial support, enhancement towards establishing and straightening green culture, and expansion of green culture. Furthermore, all data were collected via an online survey, and the findings were analysed using SPSS. According to the Spearman correlation analysis, all variables, including professional development of green entrepreneurship education, integration of sustainability into academic curricula, increase in financial support, enhancement towards establishing and straightening green culture, and expansion of green buildings and facilities, had a significant positive relationship with the presence of sustainable development of entrepreneurship education.

Previous studies support the findings. The integration of sustainability into academic curricula is an important aspect of green entrepreneurship education. According to Barrena-Martínez and Romero-Garrido (2018), integrating sustainability into academic curricula provides students with knowledge, skills, and values that are crucial for successful entrepreneurship. Moreover, the integration of sustainability into academic curricula promotes the development of critical thinking and problem-solving skills, which are important for entrepreneurship success. This suggests that the integration of sustainability into academic curricula is a key factor in the professional development of green entrepreneurship education.

Furthermore, the expansion of green buildings and facilities provides a learning environment that promotes sustainability and prepares students for sustainable entrepreneurship. According to Elgizawy (2019), sustainable facilities, such as green buildings and renewable energy systems, provide students with opportunities to learn about sustainable practices and develop sustainable business ideas. Therefore, expanding green buildings and facilities is a critical component of green entrepreneurship education.

Additionally, financial support is important for the sustainable development of green entrepreneurship education. According to Krometis and de Souza (2019), public-private partnerships can provide funding and resources needed to develop green infrastructure and facilities and promote sustainable development. This financial support can help institutions expand their green entrepreneurship education programmes and provide resources for students to develop sustainable business ideas.

Moreover, establishing and strengthening a green culture is essential for the sustainable development of entrepreneurship education. According to Alam et al. (2021), a green culture is a set of attitudes, behaviours, and practices prioritising sustainability. Establishing and strengthening a green culture can promote sustainability in entrepreneurship education, leading to the development of sustainable business ideas and practices.

In conclusion, this study suggests that a significant positive relationship exists between professional development of green entrepreneurship education, integration of sustainability into academic curricula, increase in financial support, enhancement towards establishing and strengthening green culture, expansion of green buildings and facilities, and sustainable development of entrepreneurship education. Therefore, institutions that prioritise these factors can contribute to the sustainable development of green entrepreneurship education and promote sustainability in entrepreneurship.

5.1 Contributions

The study has contributed significantly to both theory and practice in various ways. This section will discuss the contributions of this study to theory and practice. In terms of the contribution to theory, this study enhances the understanding of green entrepreneurship. The study of green entrepreneurship has contributed to the development of a theoretical understanding of the concept. This has helped clarify the characteristics, challenges, and opportunities associated with green entrepreneurship, leading to the development of a theoretical framework for green entrepreneurship. In addition, it helps in conceptualising sustainable entrepreneurship education. Sustainable entrepreneurship education has also been conceptualised, leading to the development of a theoretical framework for the integration of sustainability into academic curricula. This has helped define the critical components of sustainable entrepreneurship education and how they can be integrated into academic curricula. Moreover, it contributes to developing a framework for green entrepreneurship education, providing a comprehensive guide for institutions and educators on developing effective green entrepreneurship education programmes.

In terms of contribution to practice, the study has provided valuable insights into the development of effective green entrepreneurship education programmes, helping institutions and educators develop programmes that meet the needs of their students. The study has also contributed to integrating sustainability into academic curricula, helping institutions develop courses and programmes that provide students with knowledge, skills, and values crucial for sustainable entrepreneurship. The study has highlighted the importance of public-private partnerships in developing green entrepreneurship education programmes. This has led to the establishment of partnerships between educational institutions and private organisations, which provide the funding and resources needed to develop effective green entrepreneurship education programmes. The study has also promoted sustainable business practices by providing students with the knowledge and skills needed to develop sustainable business ideas and practices. This has led to a new generation of entrepreneurs who prioritise sustainability in their business practices.

In conclusion, the study of green entrepreneurship and sustainable entrepreneurship education development has contributed significantly to both theory and practice. It has provided valuable insights into developing effective green entrepreneurship education programmes, the integration of sustainability into academic curricula, and establishing public-private partnerships. It has also contributed to promoting sustainable business practices, leading to a new generation of entrepreneurs prioritising sustainability in their business practices.

5.2 *Implications*

This study revealed that education in the sustainable development of entrepreneurship enables every individual to acquire the knowledge, skills, attitudes, and values required to create a sustainable future. By integrating sustainability throughout entrepreneurship education, curriculum equip future leaders with the knowledge, skills, and motivation to drive meaningful action towards a sustainable future. In addition, it supports the claim that using sustainable development of entrepreneurship education as a teaching method helps students develop greater sustainability action competence. Sustainable development entrepreneurship education enables individuals to acquire the knowledge, beliefs, and skills necessary to participate in decisions regarding their actions individually and collectively, locally and globally, to improve the quality of life today while minimising potential environmental damage. Research on effectiveness can identify a number of relevant variables. It is possible to formulate long-term plans and step towards the most effective search management through research. This is essential for the researcher to make enhancements to future studies.

5.3 Limitations and Recommendations

There are several limitations of this study that can be highlighted here. In terms of population and sample size, the population of the study is limited to entrepreneurship students in one public university in Malaysia. A bigger population covering other entrepreneurship students in other institutions can be referred to in future studies. In addition, during the study, other confounding variables, such as student commitment and management commitment, may influence the outcome being studied but are not controlled for; it may be difficult to conclude the relationship between the variables of interest. Therefore, it is recommended to include these controlled variables in future studies. In terms of data analysis and method, an in-depth study is also suitable to explore the further complex relationship that can only be answered based on the interpretation by the people directly involved in the process. Instead of conducting a quantitative study, a qualitative study is another option deemed to be appropriate for achieving the research objective.

5.4 Conclusion

Independent variables such as professional development of green entrepreneurship education, integration of sustainability into academic curricula, increase in financial support, improvement towards establishing and straightening green culture, and expansion of green buildings and facilities among undergraduate students at Universiti Malaysia Kelantan were investigated. All independent variables can activate the dependent variable. Using descriptive analysis, reliability analysis, Spearman's correlation coefficient analysis, and ordinal logistics regression, data were analysed using the SPSS software. All data were collected via an online questionnaire. As a result, the excellent-to-moderate relationship between the independent and dependent variables was positively correlated with the dependent variable. The Spearman correlation analysis revealed that all independent variables had a significant positive relationship with the existence of the dependent variable, education, for the sustainable development of entrepreneurship. The ultimate objective of the study is to provide a global overview of the sustainable development of entrepreneurship education. It is significant and beneficial to society in terms of expanding our understanding of how to implement sustainable development education in their daily lives successfully.

Acknowledgements The authors would like to acknowledge ERASMUS+ CBHE ANGEL Grant No. R/ERAS/A0100/00283A/004/2021/00946; European University Cyprus, UMK Community Grant No. R/COM/A0100/00685A/002/2023/01156; UMK Prototype Grant No. R/PRO/A0100/00185A/005/2022/01131 for the support in terms of monetary and non-monetary.

References

Adegbite, S. A., Owolabi, S. A., & Akinlabi, B. H. (2020). The importance of integrating sustainability into entrepreneurship education: A systematic literature review. *Sustainability, 12*(21), 9077.

Alam, M. N., Momen, R. A., & Kabir, M. H. (2021). Entrepreneurial culture and sustainable entrepreneurship: A review. *Journal of Cleaner Production, 280*, 124321.

Al-Harthi, S. S., & Alghamdi, N. M. (2020). The impact of sustainability education on entrepreneurial intention. *Sustainability, 12*(19), 7945.

Al-Tabbaa, O., & Ankrah, S. (2019). Challenges of integrating sustainability into entrepreneurship education: A systematic review of existing literature. *Journal of Cleaner Production, 236*, 117522.

Anwar, S., & Shaari, M. S. (2021). Green infrastructure for sustainable development: A review. *Journal of Cleaner Production, 279*, 123680.

Aslani, A., & Karimi, Z. (2019). The role of government policies in promoting green entrepreneurship culture. *Journal of Cleaner Production, 236*, 117625.

Barrena-Martínez, J., & Romero-Garrido, M. I. (2018). The role of sustainability in entrepreneurship education: A systematic literature review. *Sustainability, 10*(6), 1853.

Boldureanu, G., Ionescu, A. M., Bercu, A. M., Bedrule-Grigoruță, M. V., & Boldureanu, D. (2020). Entrepreneurship education through successful entrepreneurial models in higher education institutions. *Sustainability, 12*(3), 1267.

Chen, L., Yuan, Q., & Chen, S. (2018). The impact of training on the professional development of green entrepreneurs. *Sustainability, 10*(11), 4143.

Elgizawy, S. M. (2019). The role of sustainable facilities in promoting green entrepreneurship education. *Journal of Cleaner Production, 233*, 845–857.

Fernandes, A. A. R., & Castelo Branco, M. (2021). The role of experiential learning in integrating sustainability into entrepreneurship education. *Journal of Cleaner Production, 314*, 128005.

Fu, X., & Yu, H. (2020). Policy options for expanding green infrastructure and facilities. *Journal of Cleaner Production, 277*, 123195.

Gunawan, H., & Widodo, T. (2020). The role of green entrepreneurship culture in promoting sustainable development. *Journal of Cleaner Production, 254*, 120077.

Guo, J., Li, X., & Huang, J. (2021). The importance of mentoring in the professional development of green entrepreneurs. *Journal of Business Research, 123*, 454–461.

Guzmán, C. R., & Reyes-Rodríguez, F. (2018). Effectiveness of sustainability.

Hair, J. F., Black, W. C., Babin, B. J., Anderson, R. E., & Tatham, R. L. (2006). *Multivariate data analysis*. Pearson University Press.

Kotla, B., & Bosman, L. (2023). Redefining sustainability and entrepreneurship teaching. *Trends in Higher Education, 2*(3), 498–513.

Krometis, L. A., & de Souza, M. P. (2019). Public-private partnerships for expanding green infrastructure and facilities: A review. *Journal of Cleaner Production, 220*, 839–850.

Lee, H., & Chae, J. (2018). The impact of green infrastructure on green entrepreneurship education: A case study of Seoul, South Korea. *Sustainability, 10*(6), 1955.

Li, L., Huang, J., & Wu, L. (2020). The influence of social capital on the professional development of green entrepreneurship. *Sustainability, 12*(3), 1164.

Owston, R. D. (2004). Sustaining technology innovation in the classroom: What does it take. In *Revision of paper presented at the Annual Meeting of the American Educational Research Association, San Diego*.

Potluka, O., & Svecova, L. (2019). The effects of external financial support on the capacities of educational nonprofit organizations. *Sustainability, 11*(17), 4593.

Sekaran, U., & Bougie, R. (2016). Research methods for business: A skill building approach. John Wiley & Sons.

Shahab, Y., & Rezazadeh, A. (2020). The role of social networks in promoting green entrepreneurship culture. *Journal of Cleaner Production, 272*, 122722.

Singh, A., & Jain, S. K. (2020). Incubation programs for enhancing green entrepreneurship culture: A review. *Journal of Cleaner Production, 255*, 120249.

Shulla, K., Filho, W. L., Lardjane, S., Sommer, J. H., & Borgemeister, C. (2020). Sustainable development education in the context of the 2030 Agenda for sustainable development. *International Journal of Sustainable Development & World Ecology, 27*(5), 458–468.

Wang, Y., Liu, Q., & Zhang, Y. (2020). The impact of internationalization on the professional development of green entrepreneurship. *Journal of Cleaner Production, 243*, 118475.

Wu, P., Yang, D., & Chen, H. (2020). Interactive teaching mode activates entrepreneurship classes in universities. *Open Access Library Journal, 7*(11), 1–9.

Zhang, Z., Guo, J., & Zhang, Y. (2019). The role of government policy in promoting the professional development of green entrepreneurship. *Journal of Cleaner Production, 228*, 1207–1214.

Siti Afiqah Zainuddin is a Senior Lecturer at the Faculty of Entrepreneurship and Business at Universiti Malaysia Kelantan (UMK). With over a decade of experience in teaching, she instructs students enrolled in the Bachelor of Accounting and Master of Entrepreneurship programs. Her academic qualifications include a Bachelor's Degree in Accounting (Honors) from the Islamic Science University of Malaysia, a Master's Degree in Science of Accounting from the International Islamic University of Malaysia, and a PhD in Accounting from the University of Aberdeen, Scotland, United Kingdom. She has made noteworthy contributions to SCOPUS and Web of Science-indexed publications. Additionally, she has been awarded more than fifteen research grants for various accounting and entrepreneurship projects.

Tahirah Abdullah is a senior lecturer at the Faculty of Entrepreneurship and Business, Universiti Malaysia Kelantan (UMK). She holds a Bachelor's degree in Accounting from Universiti Sains Islam Malaysia (USIM) and a Master's in Accounting from Universiti Islam Antarabangsa Malaysia (UIAM). She received her Doctor of Philosophy (PhD) in Accounting from Aberdeen, Scotland. She has been involved in many research projects and released journal articles, chapters in book and conference journals nationally and internationally, the Scopus and WOS index. Among her areas of expertise are capital management, capital management among Micro, Small, and Medium Enterprises (MSMEs), accounting in neoliberal enterprises, entrepreneurship, and education in accounting.

Amira Jamil is currently serving as a Senior Lecturer at the Faculty of Entrepreneurship and Business, Universiti Malaysia Kelantan (UMK). She earned her Doctor of Philosophy from Universiti Sultan Zainal Abidin in 2021, following a Master of Science (Accounting) in 2016 and a Bachelor of Accounting (Honors) in 2013, both from the International Islamic University Malaysia. She has published numerous academic papers and books and actively participated in various international conferences around the globe. Her areas of expertise encompass accounting, Islamic accounting, reporting, corporate governance, and taxation, with a keen interest in management, digital entrepreneurship, and digital marketing.

Nur Izzati Mohamad Anuar is a distinguished senior lecturer at the Faculty of Entrepreneurship and Business at Universiti Malaysia Kelantan. Holding a Ph.D. in Management from the University of Glasgow, Scotland, her expertise lies in organizational learning and development. With a passion for advancing the field of entrepreneurship, she has contributed significantly through her research and publications, addressing critical aspects such as knowledge, learning, and technology processes. Her work has been published in reputable journals indexed by Scopus and Web of Science, showcasing her dedication to academic excellence.

Liyana Ahmad Afip is a Faculty of Language Studies and Human Development lecturer. She is currently undertaking her PhD at the University of Queensland, Australia. Her main interest is in language-in-education policy. In addition, she is highly motivated in researching entrepreneur attributes and leadership. She has been involved in various national projects and international publications related to these topics.

Siti Rohana Mohamad is a senior lecturer in the business and accounting department, Faculty of Entrepreneurship and Business, Universiti Malaysia Kelantan (UMK). She obtained her Doctor of Philosophy from Universiti Sultan Zainal Abidin in 2023, following a Master of Science (Finance) from the International Islamic University Malaysia (IIUM) in 2013 and a Bachelor of Business Administration (Hons) (Finance) from Universiti Teknologi Mara (UiTM) in 2010. She has been involved in many research projects, presenting at conferences and publishing journal articles and books locally and abroad. Her research interests include Islamic finance, Waqf, Islamic philanthropy, social entrepreneurship, and giving behavior.

Nik Malini Nik Mahdi , since August 2012, has been a law lecturer with the Faculty of Entrepreneurship & Business of Universiti Malaysia Kelantan. She graduated and called to both the English and the Malaysian Bar. She was heavily involved, among others, in shipping and logistics negotiations vis-a-vis the commercial contracts and drafting of regulations for the State of Kelantan Islamic Foundation therein.

Muhammad Zahid Zainuddin is an Assistant Director at the Ministry of Entrepreneur Development and Cooperatives (MEDAC). With a Bachelor of Entrepreneurship (Hons) from the Northern University of Malaysia, he worked full-time toward creating a holistic and conducive entrepreneurial ecosystem to support the country's inclusive, sustainable, and driven-by knowledge and innovative entrepreneurship development agenda. In addition to being an Administrative and Diplomatic Officer (PTD) in the government sector for more than 6 years in service, he is currently entrusted with the Ministry's training programs as planning and executing the Training Operation Plan, which helps to equip sufficient knowledge and skills to cater the rapid entrepreneurship industry revolution. Close partnership and collaboration with academicians helped him explore various approaches to Government entrepreneurship's strategy and policy.

Siti Hanis Zainuddin is a Financial Accounting (Executive) at PETRONAS-EY. She obtained her Master of Islamic Finance Practice from the INCEIF University in 2023 and a Bachelor of Accounting with Honors from the Islamic Science University of Malaysia in 2017. She has been involved in many research projects and has obtained many experiences in industry.

An Empirical Study on Green Entrepreneurship and Sustainable Reporting

Siti Afiqah Zainuddin, Amira Jamil, Tahirah Abdullah, Mohd Rushdan Yasoa, Liyana Ahmad Afip, Siti Zamanira Mat Zaib, Siti Rohana Mohamad, Muhammad Zahid Zainuddin, and Siti Hanis Zainuddin

Abstract Green entrepreneurship is a crucial aspect of sustainable development goals. In exploring this discipline, the primary objective of this study is to investigate the relationship between green entrepreneurial intention and sustainable reporting. A quantitative research approach is used to explain the relationship. Data were collected from questionnaire surveys among student entrepreneurs who were run-

S. A. Zainuddin
Faculty of Entrepreneurship and Business, Universiti Malaysia Kelantan, Pengkalan Chepa, Kota Bharu, Kelantan, Malaysia

Department Accounting, Economics and Finance, School of Business Administration, European University Cyprus, Engomi, Cyprus
e-mail: sitiafiqah@umk.edu.my

A. Jamil (✉) · T. Abdullah · M. R. Yasoa · S. Z. M. Zaib · S. R. Mohamad
Faculty of Entrepreneurship and Business, Universiti Malaysia Kelantan, Pengkalan Chepa, Kota Bharu, Kelantan, Malaysia
e-mail: amira@umk.edu.my; tahirah@umk.edu.my; rushdan.y@umk.edu.my; zamanira@umk.edu.my; rohana@umk.edu.my

L. A. Afip
Faculty of Language Studies and Human Development, Universiti Malaysia Kelantan, Pengkalan Chepa, Kota Bharu, Kelantan, Malaysia
e-mail: liyana.a@umk.edu.my

M. Z. Zainuddin
Ministry of Entrepreneurship and Cooperatives Development (KUSKOP), Kuala Lumpur, Malaysia
e-mail: zahid.zainuddin@kuskop.gov.my

S. H. Zainuddin
Petroliam Nasional Berhad (PETRONAS), Kuala Lumpur, Malaysia
e-mail: sitihanis.zainuddin@petronas.com

N. Hayat, N. C. Nawi (eds.), *Contemporary Issues in Entrepreneurship and Innovative Technology*, Contributions to Environmental Sciences & Innovative Business Technology, https://doi.org/10.1007/978-3-031-50927-8_6

ning their businesses during their undergraduate study at a public university in Malaysia. For statistical analysis, correlation and descriptive analyses are conducted. Stratified random sampling is adopted for the sampling technique. The data were examined with IBM SPSS Statistics. Consequently, the vast majority of independent variables and dependent variables have a significant relationship. This study found that green product innovation, process innovation, service innovation, and technology innovation as green entrepreneurial intentions are significantly correlated to sustainability reporting.

Keywords Sustainable reporting · Green entrepreneurial intention · Student entrepreneur · Microbusiness

1 Introduction

Pengkalan Chepa was chosen for the title of the study to solicit responses from nearby residents. Numerous individuals are oblivious to the impacts of green business on sustainable reporting, and this is because most business owners are unfamiliar with green entrepreneurship. Pengkalan Chepa has many industrial areas, so selecting entrepreneurs from Pengkalan Chepa for this study is highly appropriate.

This study aims to inform the community, particularly business owners, about green entrepreneurship in greater depth. This is because they remain ignorant of the benefits of green entrepreneurship. This is because most businesses are only concerned with their bottom line. This research will demonstrate the relationship between green entrepreneurial intentions and sustainable reporting. Malaysia has introduced companies that produce eco-friendly goods and services to raise awareness of the green business. The most important phase of a product's production is the green process. This is because this level requires the implementation of specific measures to reduce the environmental impact of a product's manufacturing processes.

This study has four objectives: to determine the relationship between (i) green product innovation and sustainable reporting, (ii) green process innovation and sustainable reporting, (iii) green service innovation and sustainable reporting, and (iv) green technology innovation and sustainable reporting.

2 Literature Review

The findings from previous studies regarding the factors that contribute to sustainability reporting and disclosure are inconclusive. Therefore, adoption companies' proactive approach to green innovation reporting is reflected as stakeholders' demand for green innovation practices to be included as new practices in the green innovation component has increased. Moreover, companies value environmental reporting as part of sustainability reports because they want to demonstrate that their top management takes green issues seriously, which helps promote a positive green image.

2.1 Green Entrepreneurial Intention Variables

2.1.1 Green Product Innovation

Products are designed to minimize environmental impacts over their entire life cycle (Xie et al., 2019). Innovation in green products is a crucial factor that influences sustainable reporting. This is because a product brand that contributes to society and the environment enhances its brand image automatically. Green products are crucial in companies' efforts to achieve environmental sustainability (Pujari et al., 2003). Instantaneously, the company will have a group of loyal green consumers who prefer green products to conventional, nonenvironmentally friendly products. Even if a product contains organic ingredients, consumers will still not purchase it if it lacks a brand.

Moreover, the fabrication or production of green products will create new markets of green consumers who are even willing to pay more for the same product (Damanpour, 2010). However, consumers evaluate organic products based on green packaging, details, and other indicators that the product is green (Jaiswal & Kant, 2018). Then it will entice a greater number of stakeholders to invest in the company. Therefore, the hypothesis developed is:

H1: The relationship between green product innovation and sustainable reporting is significant.

2.1.2 Green Process Innovation

Green process innovation refers to actions that can reduce the environmental impact of all processes involved in fabricating a product. Therefore, green process development requires the incorporation of new environmentally favorable chemical pathways and technological innovation. This is because this green process innovation can enhance the organization's operational efficiency and financial performance while also fostering stakeholder confidence (Khan et al., 2021). According to Li et al. (2017), promoting green product design and production and establishing the groundwork for green product innovation is the most effective way to systematically improve the entire operation and management process to use resources and energy more efficiently. Although research is agreed to enhance company processes and reduce environmental impact, its impact on the firm's financial success remains debatable. Nevertheless, these measures can increase worker safety while minimizing negative effects such as water pollution, toxic refuse, and pollutants (Khan et al., 2021; Ma et al., 2017). Therefore, the hypothesis developed is:

H2: The relationship between green process innovation and sustainable reporting is substantial.

2.1.3 Green Service Innovation

The term "green service innovation" alludes to services that contribute to the earth's sustainability and conserve natural resources. The services in question include information technology (IT) systems, production, packaging, and storage. This can help organizations increase their operational transparency. However, each product's packaging will have a detrimental effect on the environment. This is because people are unaware of environmentally friendly packaging options. The reason is that consumers need more information before purchasing a product. Moreover, Yi et al. (2020) stated that this green service innovation could be a capital-saving resource. As less research has been conducted on green innovation variables, this will entice more academics and industries. According to Chen et al. (2015), the firm's absorptive capacity and dynamic capacity must be green, and green services must have a positive relationship with the firm's success in achieving green services. Therefore, the hypothesis developed is:

H3: The relationship between green service innovation and sustainable reporting is substantial.

2.1.4 Green Technology Innovation

Green technology innovation uses technologies such as tidal energy, vertical farming, self-sufficient structures, and natural waste treatment systems to reduce environmental impacts. The purpose of green technology is to preserve the environment, restore natural damage, and conserve the earth's natural resources. However, this has little impact on sustainable reporting because only a few entrepreneurs prioritize using green technology. The findings of a study conducted by Yamei et al. (2009) to determine the relationship between technology patents and carbon dioxide (CO2) emissions in China indicate that technological progress significantly reduces carbon emissions. In addition, Chen and Lei (2018) found a significant negative correlation between technological innovation and carbon emissions, implying that a country with a high carbon production could reduce pollution by investing more in technological innovation. Therefore, the hypothesis developed is:

H4: The relationship between green technology innovation and sustainable reporting is substantial.

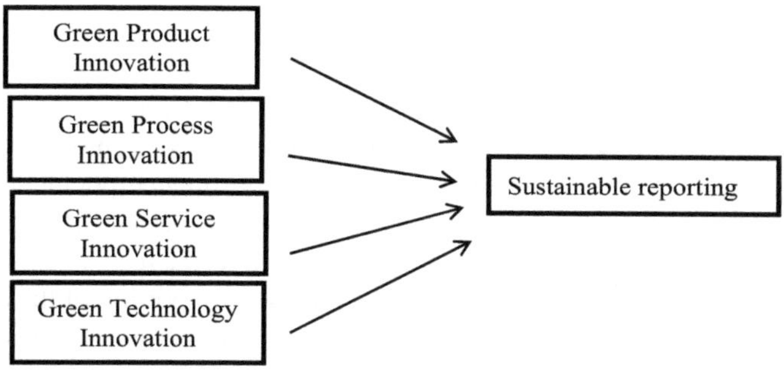

Fig. 1 Conceptual framework

2.2 Conceptual Framework

This study's objective is to examine the entrepreneurial intent of entrepreneurs in Pengkalan Chepa, Kelantan. The independent variable (IV) consists of green product innovation, process innovation, service innovation, and technology innovation, whereas the dependent variable is green entrepreneurial intent regarding sustainable reporting. The framework for the proposed research is depicted in Fig. 1.

3 Research Methodology

3.1 Research Design

Research design includes data acquisition strategies, research instruments, sampling methods, pilot testing, and data analysis techniques. This study's methodology is a quantitative research approach. Questionnaires are ideal for conducting surveys and collecting data for this study, and they contain specific questions designed to comprehend a subject from the respondents' perspective. In addition, the researcher uses secondary data as a guide to support the facts obtained from the respondents. The secondary sources utilized in this study are journal articles and Internet sources. Researchers will consult journals, articles, and reports to conduct a literature review. Journals, articles, and reports will assist the researcher in developing a framework for completing the research. The researcher utilized a variety of articles pertinent to the research topic as references to bolster the research conducted.

3.2 Research Population and Sample Size

The population is the number of individuals residing in a particular area, such as a country, state, district, etc. This study's population consists of entrepreneurs in the Pengkalan Chepa region of Kelantan, which has an estimated population of 6768 and a total area of 23 sq km. This population is the focus because it contains many entrepreneurs, making it simpler for the researcher to locate respondents to accomplish the objectives of the study. The sample represents a portion of the population to be analyzed (Arikunto, 2006). This modest data set permits the researcher to select from a large population. According to Roscoe et al. (1975), the optimal sample size for most research is between 30 and 500 individuals. Specifically, Krejcie and Morgan's sampling method determined the suitable sample size for a population of 7,000, which is 364, with an error estimate (Krejcie & Morgan, 1970). This questionnaire was distributed randomly to the respondents.

4 Analysis, Results, and Discussion

4.1 Descriptive Analysis

Table 1 to Table 4 and Fig. 2 to Fig. 5 depict the responses from the 346 respondents via surveys using the Google Forms platform. These are summaries of the demographic profiles of all respondents. Four questions were based on each participant's profile, including gender, age, race, occupation, and educational level.

Table 1 and Fig. 2 provide a comprehensive overview of the age distribution among the respondents. Predominantly, individuals aged 21 to 23 are the majority, comprising 31.6% of the sample, equivalent to 115 respondents. The subsequent age range is 24 to 26 years old, with 106 respondents representing 29.1% of the total respondents. A notable portion of the respondents, 12.9%, falls within the age range of 34 and above, with 47 respondents. Furthermore, respondent aged 27 to 29 constitute 11.3% of the sample, totaling 41 participants. The age group spanning from 18 to 20 years old comprises 10.7%, with 39 respondents. Interestingly, the

Table 1 Age

	Frequency	Percent	Valid percent	Cumulative percent
18–20 years old	39	10.7	10.7	10.7
21–23 years old	115	31.6	31.6	42.3
24–26 years old	106	29.1	29.1	71.4
27–30 years old	41	11.3	11.3	82.7
31–33 years old	16	4.4	4.4	87.1
34 years old and above	47	12.9	12.9	100.0
Total	364	100.0	100.0	

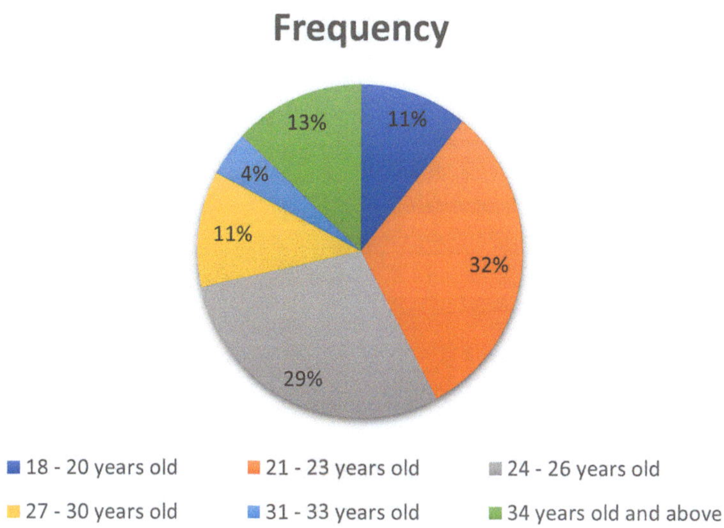

Fig. 2 Age

Table 2 Gender

Gender	Frequency	Percent	Valid percent	Cumulative percent
Male	189	51.9	51.9	51.9
Female	175	48.1	48.1	100.0
Total	364	100.0	100.0	

least represented age group is 31 to 33 years old with only 16 respondents, accounting for 4.4% of the 364 respondents.

From Table 2 and Fig. 3, the male respondents dominate female respondents from all over Pengkalan Chepa, Kota Bharu, Kelantan. One hundred eighty-nine are male respondents, representing 51.9%, while 175 are females, representing 48.1%.

Based on Table 3 and Fig. 4, most respondents are Malays, which comprises of 214 or 58.8% of the total respondents. 73 or 20.1% of the respondents are Indian, whereas 58 or 15.9% are Chinese. Respondents from other races are 19, representing 5.2% of the total respondents.

Table 4 and Fig. 5 detailed the occupational distribution of the respondents. The majority, 181 (49.7%) of the respondents are entrepreneur students. 74 respondents or 20.3% are from the private sector, whereas, 58 respondents or 15.9% are from the government sector. Notably, 47 or 12.9% of the respondents, are entrepreneurs. The remaining respondents are under the category labeled 'others' comprises a modest percentage of 1.1%, equivalent to four (4) respondents.

Based on Table 5 and Fig. 6, the majority, 196 or 53.8% of the respondents have the highest education level of undergraduate degree. 117 or 32.1% of the respondents have at least a diploma, whereas, 44 or 12.1% of the respondents are SPM-holder. The remaining four (4), or 1.1% of the respondents, have another

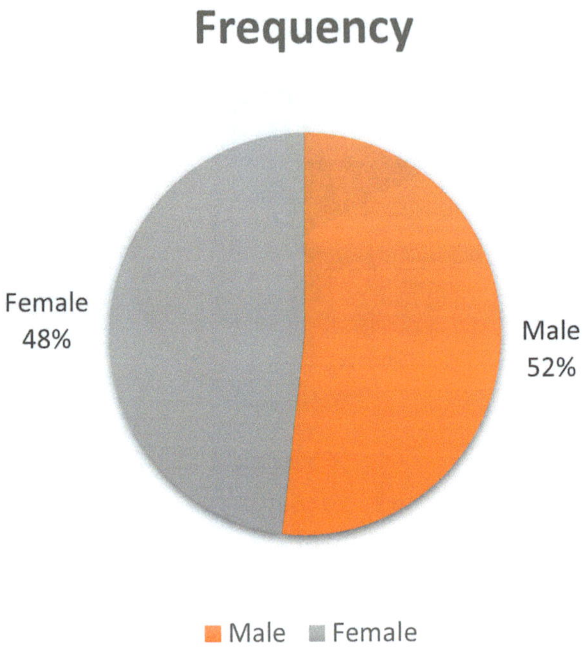

Fig. 3 Gender

Table 3 Race

Race	Frequency	Percent	Valid percent	Cumulative percent
Malay	214	58.8	58.8	58.8
Chinese	58	15.9	15.9	74.7
Indian	73	20.1	20.1	94.8
Others	19	5.2	5.2	100.0
Total	364	100.0	100.0	

education level, whereas three (3) or 0.8% of respondents have at least a master's degree.

Table 6 shows that the highest mean is green technology innovation, which is 4.3544. The second is green service innovation, with 4.3027, followed by green product innovation at 4.2445. Green process innovation has the lowest mean as an independent variable, which is 4.2319. In terms of ranking, respondents agreed that green technology innovation is the most influential factor in entrepreneurial intention related to sustainability reporting. On the other hand, green process innovation is ranked fourth by respondents.

The highest mean for this independent variable is 4.2555; the green product produced is environmentally friendly. The lowest mean for this independent variable is

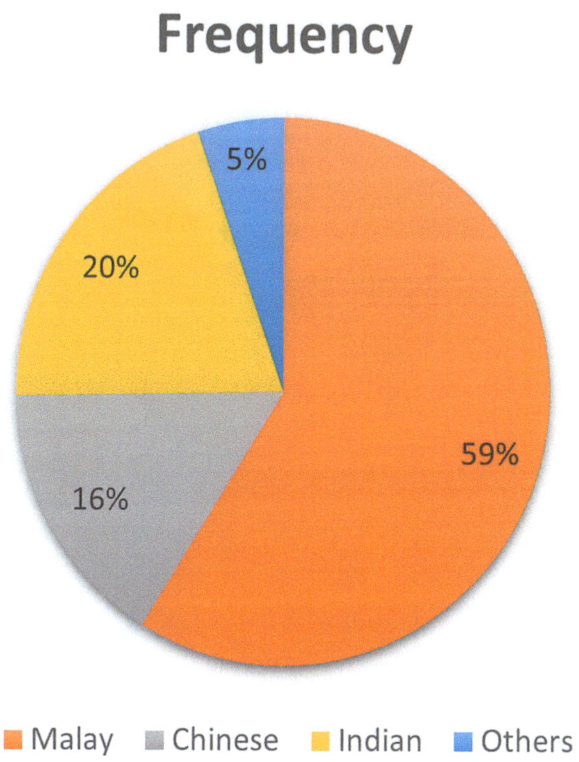

Fig. 4 Race

Table 4 Occupation

	Frequency	Percent	Valid percent	Cumulative percent
Entrepreneur	47	12.9	12.9	12.9
Entrepreneur student	181	49.7	49.7	62.6
Government sector	58	15.9	15.9	78.6
Private sector	74	20.3	20.3	98.9
Others	4	1.1	1.1	100.0
Total	364	100.0	100.0	

4.1346, that is, the green product innovation produced can reduce pollution when used (Table 7).

The highest mean for the independent variable is 4.2912, that is, green process innovation can reduce the raw materials used. On the other hand, the lowest mean for green process innovation is 4.0934, which shows that the manufacturing process of a product using green process innovation can reduce the consumption of water, electricity, coal, or oil (Table 8).

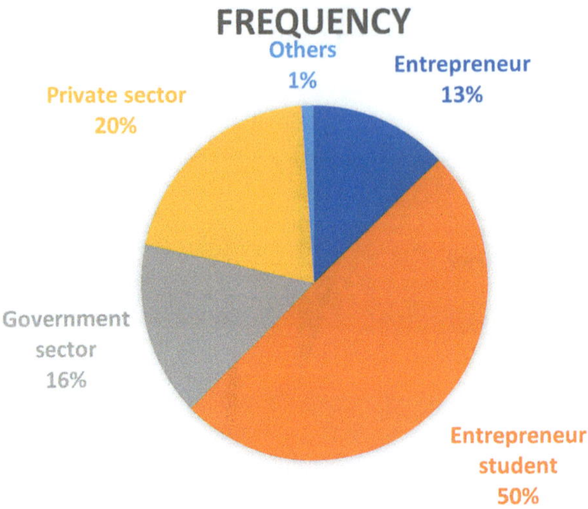

Fig. 5 Occupation

Table 5 Education

Education level	Frequency	Percent	Cumulative percent
SPM	44	12.1	12.1
Diploma	117	32.1	44.2
Undergraduate	196	53.8	98.1
Masters	3	0.8	98.9
Others	4	1.1	100.0
Total	364	100.0	

Fig. 6 Education

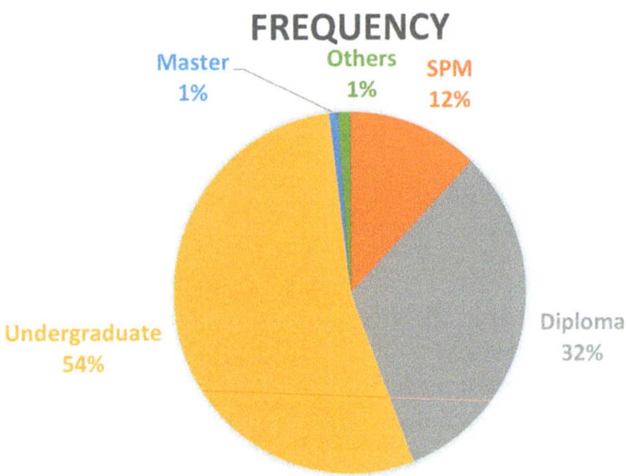

Table 6 The mean of entrepreneurial intention variables

Description	Mean
Green product innovation	4.2445
Green process innovation	4.2319
Green service innovation	4.3027
Green technology innovation	4.3544

Table 7 Green product innovation

Description	Mean
The green product innovation produced can reduce pollution when used.	4.1346
The green product innovation produced by green entrepreneurs can optimize energy and raw material resources.	4.1951
Entrepreneurs can produce green products from waste and organic materials.	4.2005
The green product produced is environmentally friendly.	4.2555
Green product is easy to recycle and reuse, so the environment is well maintained.	4.2473

Table 8 Green process innovation

Description	Mean
Green innovation processes can reduce pollution.	4.2115
Recycling waste materials can be done to keep the environment in good condition.	4.2775
The process of reducing prohibited substances can maintain a good environment.	4.2857
Green process innovation can reduce the use of raw materials.	4.2912
The manufacturing process of a product using green process innovation can reduce the consumption of water, electricity, coal, or oil.	4.0934

Table 9 Green service innovation

Description	Mean
Green service innovation can reduce dependence on nonrenewable energy sources.	4.2115
Green service innovation can protect ecosystem balance.	4.2775
Green service innovation makes us more socially responsible toward the environment.	4.2857
Green service innovation can improve air quality and community life.	4.2912
Green service innovation affects healthy competitive advantage.	4.0934

The highest mean score goes to green service innovation can improve air quality and community life, with 4.2912. The lowest mean is green service innovation affects healthy competitive advantage, with 4.0934 (Table 9).

The highest mean score for this independent variable is 4.3626, which represents green technology innovation as being safe to use in the long term. The lowest mean score goes to green technology innovation is hard to come by because not everyone understands the concept of “green entrepreneurship,” which is 4.2143 (Table 10).

Table 10 Green technology innovation

Description	Mean
Green technology innovation is hard to come by because not everyone understands the concept of "green entrepreneurship."	4.2143
Many types of green innovation technologies can be used to reduce the impact on the environment.	4.3022
This green technology innovation is safe to use in the long term.	4.3626
Green technology innovation can explore opportunities to export green products that have been produced.	4.3187
Green technology innovation can reduce carbon emissions into the air.	4.3159

Table 11 Results of Pearson correlation coefficient

	Green production	Green process	Green services	Technology innovation	Sustainable reporting
Green production	1	0.823** (0.000)	0.802** (0.000)	0.769** (0.000)	0.668** (0.000)
Green process	0.823** (0.000)	1	0.832** (0.000)	0.775** (0.000)	0.652** (0.000)
Green services	0.802** (0.000)	0.832** (0.000)	1	0.849** (0.000)	0.666** (0.000)
Technology innovation	0.769** (0.000)	0.775** (0.000)	0.849** (0.000)	1	0.688** (0.000)
Sustainable reporting	0.668** (0.000)	0.652** (0.000)	0.666** (0.000)	0.688** (0.000)	1

**Correlation is significant at the 0.01 level (two tailed)

4.2 *Pearson's Correlation Coefficient*

Table 11 shows the result of Pearson's correlation coefficient. Pearson's correlation coefficient was used to test the hypotheses. It shows correlation measurements to evaluate the strength and direction of linear and two-variable relationships. All variables are said to be significantly positively correlated when a positive change in the other variable accompanies a change in one variable. These indicate that all of the hypotheses are accepted. In other words, the green entrepreneurial intention factor represented by green products, processes, services, and technology innovation positively correlates to sustainability reporting.

5 Discussions, Recommendations, and Conclusion

The study found that firms engaged in green products, processes, services, and technology innovation were more likely to report on their sustainability performance. This suggests that firms committed to environmental sustainability are more likely to communicate their efforts to stakeholders through sustainability reporting.

Moreover, the study found that the level of sustainability reporting was positively related to the level of green entrepreneurship. This suggests that firms engaging in green entrepreneurship are more likely to be transparent about their sustainability performance and communicate their sustainability efforts to stakeholders. The main implication of this study is green product innovation, process innovation, service innovation, and technology innovation. In addition, the researcher also focuses on sustainable reporting. The result has shown that the correlation for green product innovation is the highest compared to the other two independent variables. In addition, the significance of green product innovation should not be underestimated as it can affect sustainable reporting. Finally, in the context of research, green technology innovation is lacking, as shown in a moderate correlation. This can conclude that student entrepreneur acceptance of green technology innovation is still low and can be increased with certain improvements.

There are several limitations and recommendations for future researchers. First, the current study only covers the population in east coast Malaysia. In the future, researchers can conduct studies in broader areas around Malaysia to enrich the local context. Future researchers may seek more information about the effect of green entrepreneurial intention on sustainable reporting. Besides that, future researchers can find out how important green products, services, technology, and processes are. Second, the research period is too short, that is, 3 months. Future researchers are advised to devote a longer period for data collection so they can have enough time to find people who are more suitable and qualified to be respondents. It is challenging to finish the procedure of gathering data for the sample size in a short amount of time.

Next, the researcher should prepare questions that are easy for respondents to understand. This is very important to ensure that respondents are able to answer the questions efficiently. For researchers to obtain more detailed and accurate information, the respondents should have clarity and better understanding.

The study has contributed significantly to both theory and practice in various ways. In terms of the contribution to theory, this study enhances the understanding of green entrepreneurship. The study of green entrepreneurship has contributed to developing a theoretical understanding of the concept. This has helped clarify the characteristics, challenges, and opportunities associated with green entrepreneurship, leading to the development of a theoretical framework for green entrepreneurship. In addition, it helps in conceptualizing green entrepreneurial intention and sustainable reporting. This has helped define the critical components of green entrepreneurial intention and how they can be integrated into sustainability reporting. Moreover, it contributes to developing a framework for green entrepreneurship intention, which provides a comprehensive guide for business owners on developing effective firm sustainability reporting.

In terms of contribution to practice, the study has provided valuable insights into developing effective sustainability reporting, helping business owners develop a report representing the suitable and relevant components of green entrepreneurship. The study has also contributed to the integration of green entrepreneurial intention elements into sustainability reports, helping businesses develop relevant and reflective reports that provide stakeholders with knowledge, understanding, and values crucial for sustainable entrepreneurship development.

Acknowledgments The authors would like to acknowledge ERASMUS+ CBHE ANGEL Grant No. R/ERAS/A0100/00283A/004/2021/00946; European University Cyprus, UMK Community Grant No. R/COM/A0100/00685A/002/2023/01156; and UMK Prototype Grant No. R/PRO/A0100/00185A/005/2022/01131 for the support in terms of monetary and nonmonetary.

References

Arikunto, S. (2006). Prosedur Penelitian Suatu Pendekatan Praktik. *Jakarta. Jurnal Gerbang, 8*(1).

Chen, W., & Lei, Y. (2018). The impacts of renewable energy and technological innovation on environment-energy-growth nexus: New evidence from a panel quantile regression. *Renewable energy, 123*, 1–14.

Chen, Y. S., Lin, Y. H., Lin, C. Y., & Chang, C. W. (2015). Enhancing green absorptive capacity, green dynamic capacities and green service innovation to improve firm performance: An analysis of structural equation modeling (SEM). *Sustainability, 7*(11), 15674–15692.

Damanpour, F. (2010). An integration of research findings of effects of firm size and market competition on product and process innovations. *British Journal of Management, 21*(4), 996–1010.

Jaiswal, D., & Kant, R. (2018). Green purchasing behaviour: A conceptual framework and empirical investigation of Indian consumers. *Journal of Retailing and Consumer Services, 41*, 60–69.

Khan, P. A., Singh, S. K. J. P., Johl, S. K., Shamim, A., Nurhayadi, Y., Wijiharjono, N., & Al-Azizah, U. S. (2021). Injecting green innovation reporting into sustainability reporting. *SHS Web of Conferences*.

Krejcie, R. V., & Morgan, D. W. (1970). Determining sample size for research activities. *Educational and psychological measurement, 30*(3), 607–610.

Li, D., et al. (2017). The impact of legitimacy pressure and corporate profitability on green innovation: Evidence from China top 100. *Journal of Cleaner Production, 141*, 41–49.

Ma, Y., et al. (2017). Green process innovation and innovation benefit: The mediating effect of the firm image. *Sustainability, 9*(10), 1778.

Pujari, D., Wright, G., & Peattie, K. (2003). Green and competitive: Influences on environmental new product development performance. *Journal of business Research, 56*(8), 657–671.

Roscoe, A. M., Lang, D., & Sheth, J. N. (1975). Follow-up methods, questionnaire length, and market differences in mail surveys: In this experimental test, a telephone reminder produced the best response rate and questionnaire length had no effect on rate of return. *Journal of Marketing, 39*(2), 20–27.

Xie, X., et al. (2019). Green process innovation, green product innovation, and corporate financial performance: A content analysis method. *Journal of Business Research, 101*, 697–706.

Yamei, H., Zheng, M., Zhengtao, L., & Lirong, G. A. O. (2009). Distribution of polycyclic aromatic hydrocarbons in sediments from Yellow River Estuary and Yangtze River Estuary, China. *Journal of Environmental Sciences, 21*(12), 1625–1631.

Yi, M., Wang, Y., Sheng, M., Sharp, B., & Zhang, Y. (2020). Effects of heterogeneous technological progress on haze pollution: Evidence from China. *Ecological Economics, 169*, 106533.

Siti Afiqah Zainuddin is a Senior Lecturer at the Faculty of Entrepreneurship and Business at Universiti Malaysia Kelantan (UMK). With over a decade of experience in teaching, she instructs students enrolled in the Bachelor of Accounting and Master of Entrepreneurship programs. Her academic qualifications include a Bachelor's Degree in Accounting (Honors) from the Islamic Science University of Malaysia, a Master's Degree in Science of Accounting from the International

Islamic University of Malaysia, and a PhD in Accounting from the University of Aberdeen, Scotland, United Kingdom. She has made noteworthy contributions to SCOPUS and Web of Science-indexed publications. Additionally, she has been awarded more than 15 research grants for various accounting and entrepreneurship projects.

Amira Jamil is currently serving as a Senior Lecturer at the Faculty of Entrepreneurship and Business, Universiti Malaysia Kelantan (UMK). She earned her Doctor of Philosophy from Universiti Sultan Zainal Abidin in 2021, following a Master of Science (Accounting) in 2016 and a Bachelor of Accounting (Honors) in 2013, both from the International Islamic University Malaysia. She has published numerous academic papers and books and actively participated in various international conferences around the globe. Her areas of expertise encompass accounting, Islamic accounting, reporting, corporate governance, and taxation, with a keen interest in management, digital entrepreneurship, and digital marketing.

Tahirah Abdullah is a senior lecturer at the Faculty of Entrepreneurship and Business, Universiti Malaysia Kelantan (UMK). She holds a Bachelor's degree in Accounting from Universiti Sains Islam Malaysia (USIM) and a Master's in Accounting from Universiti Islam Antarabangsa Malaysia (UIAM). She received her Doctor of Philosophy (PhD) in Accounting from Aberdeen, Scotland. She has been involved in many research projects and released journal articles, chapters in book and conference journals nationally and internationally, the Scopus and WOS index. Among her areas of expertise are capital management, capital management among Micro, Small, and Medium Enterprises (MSMEs), accounting in neoliberal enterprises, entrepreneurship, and education in accounting.

Mohd Rushdan Yasoa ' is a senior lecturer at the Faculty of Entrepreneurship and Business, Universiti Malaysia Kelantan (UMK). He holds a Bachelor in Accountancy (with honors) from Universiti Kebangsaan Malaysia (UKM) and a Master's in Business Administration (Accounting) from Universiti Utara Malaysia (UUM). He received his Doctor of Philosophy (PhD) in Accounting from Universiti Sultan Zainal Abidin (UniSZA). He has been involved in many research projects and published articles in journals, book chapters, and conference journals nationally and internationally that Scopus and Wos index. Her areas of expertise are Shariah Audit, Accounting Information Systems, Financial Management in Small and Medium Enterprises (SMEs) and religious institutions, and accounting education. Apart from being an academic scholar, he is also an associate member of the Malaysian Institute of Accountants (MIA).

Liyana Ahmad Afip is a Faculty of Language Studies and Human Development lecturer. She is currently undertaking her PhD at the University of Queensland, Australia. Her main interest is in language-in-education policy. In addition, she is highly motivated in researching entrepreneur attributes and leadership. She has been involved in various national projects and international publications related to these topics.

Siti Zamanira Mat Zaib is a lecturer at the Faculty of Entrepreneurship and Business, Universiti Malaysia Kelantan (UMK). She holds a Bachelor's degree in Fiqh and Fatwa and a Master's Degree in Muamalat Administration and Halal Product, both from Universiti Sains Islam Malaysia (USIM). She obtained her Doctor of Philosophy (PhD) in Islamic Banking and Finance from Universiti Malaysia Kelantan (UMK). She has published her research project academically, including journal chapters in the book, and actively participated in various conference journals nationally and internationally. Among her areas of expertise are halal studies, including halal food and halal product management, halal financing, and Islamic banking and finance areas.

Siti Rohana Mohamad is a senior lecturer in the business and accounting department, Faculty of Entrepreneurship and Business, Universiti Malaysia Kelantan (UMK). She obtained her Doctor of Philosophy from Universiti Sultan Zainal Abidin in 2023, following a Master of Science (Finance) from the International Islamic University Malaysia (IIUM) in 2013 and a Bachelor of Business Administration (Hons) (Finance) from Universiti Teknologi Mara (UiTM) in 2010. She has been involved in many research projects, presenting at conferences and publishing journal articles and books locally and abroad. Her research interests include Islamic finance, Waqf, Islamic philanthropy, social entrepreneurship, and giving behavior.

Muhammad Zahid Zainuddin is an Assistant Director at the Ministry of Entrepreneur Development and Cooperatives (MEDAC). With a Bachelor of Entrepreneurship (Hons) from the Northern University of Malaysia, he worked full-time toward creating a holistic and conducive entrepreneurial ecosystem to support the country's inclusive, sustainable, and driven-by knowledge and innovative entrepreneurship development agenda. In addition to being an Administrative and Diplomatic Officer (PTD) in the government sector for more than 6 years in service, he is currently entrusted with the Ministry's training programs as planning and executing the Training Operation Plan, which helps to equip sufficient knowledge and skills to cater the rapid entrepreneurship industry revolution. Close partnership and collaboration with academicians helped him explore various approaches to Government entrepreneurship's strategy and policy.

Siti Hanis Zainuddin is a Financial Accounting (Executive) at PETRONAS-EY. She obtained her Master of Islamic Finance Practice from the INCEIF University in 2023 and a Bachelor of Accounting with Honors from the Islamic Science University of Malaysia in 2017. She has been involved in many research projects and has obtained many experiences in industry.

Behavioral Intention to Use ICT Among Micro-Entrepreneurs

Tahirah Abdullah, Siti Afiqah Zainuddin, Amira Jamil, Siti Salwani Abdullah, Nur Farahiah Azmi, Nur Izzati Mohamad Anuar, Nik Alif Amri Nik Hashim, and Siti Rohana Mohamad

Abstract The present study aims to explore the behavioral intention of utilizing information and communication technology (ICT), which is considered a vital component of organizational strategies for small and medium-sized enterprises (SMEs). The widespread use of ICT has become indispensable in modern life, contributing significantly to business performance, growth, and competitiveness. However, microenterprises face challenges in fully leveraging the potential benefits of ICT due to their limited understanding and skills in managing and implementing these technologies. To address this issue, the study investigates the barriers to effective ICT use in microenterprises and identifies ways to enhance their performance. The questionnaire is distributed to microenterprise owners, consisting of students who have established their own businesses under Universiti Malaysia Kelantan Entrepreneurship Institute. The study recommends expanding the use of ICT in business and developing a comprehensive understanding of ICT in business to improve microenterprise performance and keep pace with cutting-edge technology.

T. Abdullah · A. Jamil · S. S. Abdullah · N. F. Azmi · N. I. M. Anuar · S. R. Mohamad
Faculty of Entrepreneurship and Business, Universiti Malaysia Kelantan,
Pengkalan Chepa, Kota Bharu, Kelantan, Malaysia
e-mail: tahirah@umk.edu.my; amira@umk.edu.my; salwani.a@umk.edu.my; nurfarahiah@umk.edu.my; nurizzati.ma@umk.edu.my; rohana@umk.edu.my

S. A. Zainuddin (✉)
Faculty of Entrepreneurship and Business, Universiti Malaysia Kelantan,
Pengkalan Chepa, Kota Bharu, Kelantan, Malaysia

Department Accounting, Economics and Finance, School of Business Administration,
European University Cyprus, Engomi, Cyprus
e-mail: sitiafiqah@umk.edu.my

Nik Alif Amri Nik Hashim
Faculty of Hospitality, Tourism and Wellness, Universiti Malaysia Kelantan,
Pengkalan Chepa, Kota Bharu, Kelantan, Malaysia
e-mail: alifamri@umk.edu.my

N. Hayat, N. C. Nawi (eds.), *Contemporary Issues in Entrepreneurship and Innovative Technology*, Contributions to Environmental Sciences & Innovative Business Technology, https://doi.org/10.1007/978-3-031-50927-8_7

Keywords Intention · Information and communication technology (ICT) · SMEs · Microenterprise

1 Introduction

Information and communication technology (ICT) has become integral to modern human life. Modern life would be unimaginable without the pervasiveness of ICT. It is commonly acknowledged that small and medium enterprises (SMEs) organizational plans must now include the implementation of ICTs. That is a fundamental source of promoting innovation and providing performance, growth, and competitiveness for companies able to leverage these technologies (Busaidi et al., 2019). Many developing nations are attempting to use ICT to create SMEs, given the success of the technology in SMEs in developed nations. People, businesses, and society as a whole have been using ICT in unprecedented amounts for a variety of purposes. There are four types of businesses in Malaysia which are micro, small, medium, and large. SMEs are also considered an important engine of economic growth in each country.

Microenterprise is still growing all around the world (Pedraza, 2021). These companies, which are described as small, owner-operated firms that people from underrepresented groups frequently found, adopt different organizational structures to deal with entrance restrictions and financial limitations (Pedraza, 2021). A microenterprise in Malaysia typically has fewer than ten people and very little capital (Sandberg & Håkansson, 2020). Most microenterprises focus on offering local products or services (Sandberg & Håkansson, 2020). These microenterprises can operate out of homes, on farms, or on the streets, and they can also operate part- or full-time (Sandberg & Håkansson, 2020). In both established and developing nations, microenterprises are prevalent in all areas of business creation, including rural, urban, and suburban areas (Yunis et al., 2018). Small businesses must make effective use of information and communication technology. It primarily aids in the generation of more profit from the creation of novel business prospects, as well as the development of a competitive edge. In order to capitalize on the commercial opportunities provided by ICT and the innovations that result from it, micro-entrepreneurs' thoughts and actions are necessary. Microenterprise owners must thus be fluent in the language of technology to connect technical potential with changing market conditions, new client demands, growing issues, and prospective business prospects (Yunis et al., 2018).

Effective ICT use is crucial for the viability of tiny businesses (Omar, 2018). The solution to this new difficult problem will thus have far-reaching effects on how firms approach IT investment and management. The use of ICT relies on e-commerce to generate microenterprise performance (Omar, 2018). Utilizing a resource-based approach, integrating ICT with front-end features and back-end databases can result in special ICT capabilities that are difficult to duplicate, potentially enhancing the business performance of microenterprises (Kwabiah, 2019). According to Mwenda

(2018), the influence of this ICT use component, "ICT integration," on microenterprise performance has not been clearly defined since it has not been well documented in recent study literature. Similarly, even when ICT is utilized appropriately and effectively, "ICT utilization" is another facet of ICT use that may be important to company success.

A key obstacle to adopting ICT in microenterprises is the lack of awareness about the potential advantages of ICT and the strategies that can aid them in achieving their objectives (Orser et al., 2019). Microenterprises face challenges as they typically handle their own ICT and make nearly all business decisions independently. Unfortunately, the limitations of the owner or manager can hinder the company's ability to perform specific tasks effectively, leading to inefficiencies.

In addition, the belief that technology does not suit their specific business nature and demands is a significant factor in why microenterprises do not intend to use ICT for business purposes (Pillay, 2016). ICT unsuitability for business also has significant implications for SMEs that wish to employ the technology since they will not do so until it is absolutely necessary to create and maintain a more complex ICT-based system. As a result, most likely have a negative view of ICT use, which also prevents the majority of microenterprises from engaging in ICT activities and links the behavioral intention of microenterprises to adopt ICT use. However, with effective adoption of ICT, the organization can attempt to develop a better performance of the microenterprises while at the same time making them intend to adopt the use of ICT in doing their business. These issues have led to the following objectives:

1. To examine the relationship between the perceived usefulness and behavioral intention to use ICT among microenterprise.
2. To examine the relationship between the perceived ease of use and behavioral intention to use ICT among microenterprise.
3. To examine the relationship between social influence and behavioral intention to use ICT among microenterprise.
4. To examine the relationship between the facilitating conditions and behavioral intention to use ICT among microenterprise.

2 Literature Review

This study examines the behavioral intention to use ICT among microenterprises in a public university in Kelantan. The theory of acceptance model (TAM) was developed by Davis in 1986 as a modification of the theory of reasoned action (TRA). This theory examined the acceptance of ICT technology and microenterprise performance among students who owned micro businesses. The goal of TAM is to explain and predict the extent to which owners of enterprises will adopt new technology (Davis, 1989). Perceived usefulness (PU) and perceived ease of use (PEOU) are the most important characteristics that determine behavioral intention, directly or indirectly.

In addition, the unified theory of acceptance and use of technology model (UTAUT) by Venkatesh et al. (2003) was also used to frame the present study. This model extends the TAM (Rozmi et al., 2019). The degree to which a person feels that important individuals believe they should utilize the new system is characterized as a social influence (Dwivedi et al., 2019). Several studies have identified a relationship between social influence and behavioral intention to use ICT in developed countries and new industrial countries (Hariri, 2014; Yueh et al., 2015). Furthermore, facilitating conditions are described as an individual's belief that a technological and organizational framework is available to allow the system to be used (Dwivedi et al., 2019). According to Handoko (2019), the independent variable of facilitating conditions had no significant influence on behavioral intention, but it did significantly impact ICT use. As a result, the UTAUT model was designed only to see the effect of facilitating conditions.

Besides, PU can measure how much a technology adopter use of it can improve the efficiency of their organizational performance (Suhartanto & Leo, 2018). This term, known as a utilitarian outcome, is important in the acceptability of technology in the microenterprise. The utilitarian outcome is important because the high value of ICT will make microenterprise technology more attractive. Indirectly, the PU of ICT can increase the desire to adopt ICT usage among the owner of microenterprises. Next, PEOU is a person's belief that the system is simple to understand, and the intensity of use and interaction between users and the system can make this ICT more convenient to use (Zuniarti et al., 2021). Furthermore, PEOU is a variable used to assess how much effort a person believes it takes to use the system (Gillies et al., 2020). PEOU is a basic concept that affects a system's acceptability together with perceived utility. A number of studies have recognized its importance in the usage of ICT.

Social influence is the growth of a person's beliefs, emotions, mentalities, and behavior as a result of interactions with other people or groups (Rahman et al., 2021). Therefore, social influence refers to the impact of other people's beliefs and actions, such as coworkers, peers, and families, on a person's behavior. In the context of wearable ICT, a person's behavioral intention toward ICT is affected by their environment (Rahman et al., 2021). Moreover, facilitating condition is the level where each user feels that the organization's infrastructure and technology are available to facilitate ICT users (Venkatesh et al., 2016).

Behavioral intention to use can be characterized as either a purchasing signal or a diagnostic value to use new technology (Baig & Shahzad, 2022). According to Tuncer et al. (2021), behavioral intentions refer to a customer's willingness to use the service again, suggest to others, and share positive words about the business. Behavioral intention is a significant indicator of whether a user would stay with or abandon a certain service on ICT (Trivedi & Yadav, 2020).

3 The Research Model and Hypothesis Development

The present study uses the theory of acceptance model (TAM) and unified theory of acceptance and use of technology model (UTAUT) as a base model to investigate microenterprises' behavioral intention to use ICT in their business entities. The study considers perceived usefulness, perceived ease of use, social influence, and facilitating condition as determinants toward behavioral intention to use ICT among microenterprises. Figure 1 shows the research model.

Perceived usefulness can measure how much a technology adopter's use of it can improve the efficiency of their organizational performance (Suhartanto & Leo, 2018). This term, known as a utilitarian outcome, is important in the acceptability of technology in the microenterprise. Suhartanto and Leo (2018) proclaimed that perceived usefulness refers to a person's perception that a certain technology individual is considering adopting would improve their potential to perform their work. This is important to learn because the high value of ICT will make microenterprise technology more attractive. Indirectly, the perceived usefulness of ICT can increase the desire for adoption among the microenterprise. Previous studies by Camilleri (2020) revealed that there is a significant relationship between the perceived usefulness of behavioral intention to use ICT, mainly electronic media, and advanced innovative technology among enterprise entities. Thus:

H_1: Perceived usefulness positively affects the behavioral intention to use ICT among microenterprises in Universiti Malaysia Kelantan.

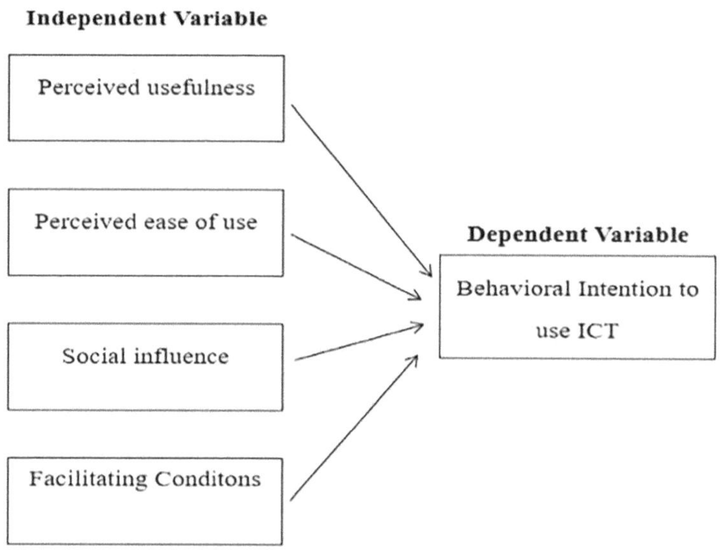

Fig. 1 Research framework

The belief that a system is easy to understand is known as perceived ease of use. The intensity of usage and user-system interaction can enhance the ICT's convenience (Basuki et al., 2022). Basuki et al. (2022) established that perceived ease of use (PEOU) is a practical variable for assessing the simplicity of using ICT. PEOU also reflects the level of effort an individual believes is required to use a system (Huang et al., 2022) and is often considered a crucial determinant of attitudes and perceived usefulness (PU). PEOU is a fundamental concept that influences a system's acceptability, along with perceived utility, and thus:

H_2: Perceived ease of use positively affects the behavioral intention to use ICT among microenterprises in Universiti Malaysia Kelantan.

Social influence is the growth of a person's beliefs, emotions, mentalities, and behavior as a result of interactions with other people or groups (Rahman et al., 2021). In other words, it impacts other people's beliefs and actions, such as coworkers, peers, and families, on a person's behavior. In the context of wearable ICT, a person's behavioral intention toward ICT is affected by their environment (Rahman et al., 2021). The degree to which ICT users believe that those relevant to them believe they need to use ICT is referred to as social influence. Social influence is a significant factor influencing a system's acceptance (Subawa et al., 2020). Khazaei (2020) conducted a study on the impact of ICT on society and found that social influence can predict the use of ICT, whereby the users of ICT among microenterprises have less and lower levels of ICT use, while ICT users in medium and large enterprises have higher levels of ICT usage.

H_3: Social influence positively influences the behavioral intention to use ICT among microenterprises in Universiti Malaysia Kelantan.

Facilitating condition is the level where each user feels that the organization's infrastructure and technology are available to facilitate ICT users in the organization (Venkatesh et al., 2016). Besides, the setting where management and technical arrangements exist to enable the use of systems that facilitate access to the resources and knowledge required to operate the system also facilitates the intention to use technology. According to Gupta et al. (2020), there are many dimensions to facilitate the situation in the study conducted; among them are compatibility, required resources, expertise, specific people, and the help of others. Therefore, the ease of using ICT affects the behavior of using ICT in business. This is because using ICT in organizational infrastructure can facilitate more sophisticated business operations. Ambarwati et al. (2020) discovered that facilitating conditions significantly positively influenced behavioral intention to utilize the technological infrastructure. Certain wearable technology may have complex application-level interfaces, and others may present more straightforward ways to use the device. Thus:

H_4: Facilitating condition positively affects the behavioral intention to use ICT among microenterprises in Universiti Malaysia Kelantan.

4 Methodology

The study was conducted to understand the behavioral intention to use ICT among microenterprises in UMK. Thus, research framework was developed to predict behavioral intention to use ICT among microenterprises in their companies. Primary data were collected through a questionnaire. The questionnaires were designed to gather information about the research model variables among the respondents in Universiti Malaysia Kelantan. A Likert scale is appropriate when the research needs to measure the respondents' attitudes toward constructs. Therefore, a five-point Likert scale (1 = strongly disagree to 5 = strongly agree) was designed for this study to collect the information. The present study chooses the owner and manager of microenterprises because they are the company's most influential decision-makers (Damanpour & Schneider, 2006). The population for this study is the students from the Faculty of Business Entrepreneurship and Faculty of Hospitality, Tourism, and Wellness, Universiti Malaysia Kelantan, which is located at Pengkalan Chepa, Kota Bharu, Kelantan. A total of 310 questionnaires were distributed to the owner of microenterprises, consisting of students from both faculties. All of them completed the questionnaire with a response rate of 100%.

5 Profile of the Respondent

A total of 310 respondents completed the survey, of which 34.2% were male and 65.8% were female (Table 1). Regarding age, most of the respondents, which were 69%, were between 23 and 24 years old and owned and managed a business. The majority of the respondents were at year 4 which were 68.1%, and the least were year 1 students, representing 3.9% of the total. Around 5% of the respondents earned business income between RM2000 and RM2999, most of which were 69% earning below RM1000 monthly income.

5.1 Data Analysis and Discussion

Cronbach's alpha was used to evaluate the data's reliability to a common internal accuracy measure (Chan et al., 2010). The reliability is completely reflected if the alpha value is 1. According to George and Mallery, Cronbach's alpha values above 0.90 have excellent outstanding internal consistency, above 0.80 are good, and above 0.70 are acceptable (Masadeh et al., 2016). Table 2 shows the value of cronbach's alpha are above 0.7 for all items. Items score above 0.7 indicates strong internal consistency within the measure. This means that the items are highly consistent in measuring the same underlying construct.

Table 1 Profile of the respondents

Characteristics	Number	Percentage (%)
Gender		
Male	106	34.2
Female	204	65.8
Total	310	100
Race		
Malay	241	77.7
Chinese	42	13.5
Indian	24	7.7
Others	3	1
Total	310	100
Age		
19–20 years	15	4.8
21–22 years	62	20
23–24 years	214	69
Above 25 years	19	6.1
Total	310	100
Year of study		
Year 1	12	3.9
Year 2	32	10.3
Year 3	55	17.7
Year 4	211	68.1
Total	310	100
Faculty		
FKP	262	84.5
FHPK	48	15.5
Total	310	100
Course		
SAK	134	43.2
SAL	29	9.4
SAB	29	9.4
SAR	44	14.2
SAE	17	5.5
SAA	9	2.9
SAP	13	4.2
SAH	14	4.5
SAS	21	6.8
Total	310	100
Monthly income		
Below RM1000	214	69
RM1000–RM1999	81	26.1
RM2000–RM2999	15	4.8
RM3000–RM3999	0	0
Above RM4000	0	0
Total	310	100

Table 2 Summary of reliability test

Construct	Cronbach's alpha
Perceived usefulness	0.722
Perceived ease of use	0.738
Social influence	0.779
Facilitating condition	0.750
Behavioral intention to use ICT	0.775

Table 3 Coefficient analysis

Variable	Unstandardized coefficients	Std error	*T*-value	Sig.	VIF
Perceived usefulness	0.278	0.063	4.681	0.000	2.585
Perceived ease of use	0.085	0.072	1.171	0.243	3.839
Social influence	0.338	0.065	4.990	0.000	3.350
Facilitating condition	0.144	0.063	2.265	0.024	2.945

Dependent variable. Intention to use ICT

Based on Table 3, it has shown the value of the coefficient of multiple linear regression. From the results obtained, independent variables of perceived usefulness, social influence, and facilitating condition are significant because their *p*-values are less than 0.05. However, variables perceived ease of use are not significant because the *p*-value of it is greater than 0.05.

The unstandardized coefficients beta describes how strongly each independent variable influences the dependent variable. The greater the value of the beta coefficient, the stronger the effect of the independent variables on the dependent variable. According to Table 3, the most influential variable in measuring behavioral intention to use ICT among microenterprises is social influence ($\beta = 0.324$), and the second is perceived usefulness ($\beta = 0.293$). Third is facilitating condition ($\beta = 0.142$); the least influential variable is perceived ease of use ($\beta = 0.084$). Since social influence is the most influencing variable and its definition refers to the extent to which the belief is either positive or negative, other people, especially family, friends, and peers, will influence the choice of an individual to accept the new system; the respondents in this study agree that it is their contacts that influence their behavioral intention to use ICT in their company.

Table 4 confirms the hypothesis stating that perceived usefulness, perceived ease of use, social influence, and facilitating conditions of technology influence the intention to use ICT among microenterprise owners and managers.

6 Findings and Discussions

The findings suggest that three hypotheses were accepted. These factors confirmed that perceived usefulness, social influence, and facilitating conditions of technology influence the intention to use ICT among microenterprise owners and managers.

Table 4 The confirmation of the hypotheses

Hypotheses	Result
H1: Perceived usefulness positively affects the behavioral intention to use ICT among microenterprises in Universiti Malaysia Kelantan	Accepted
H2: Perceived ease of use positively affects the behavioral intention to use ICT among microenterprises in Universiti Malaysia Kelantan	Rejected
H3: Social influence positively influences the behavioral intention to use ICT among microenterprises in Universiti Malaysia Kelantan	Accepted
H4: Facilitating condition positively impacts the behavioral intention to use ICT among microenterprises in Universiti Malaysia Kelantan	Accepted

The perceived usefulness was positively significant because when the owner of the microenterprises perceived the usefulness of using the ICT, it increased their intention to use the technology in their business operations. This was consistent with the findings of the studies from Buabeng-Andoh (2018) and Suhartanto and Leo (2018), which conclude that perceived usefulness relates to the level to which an individual feels technology is valuable and the extent to which the usage of the technology can increase the performance efficiency of their organization.

In addition, social influence and behavioral intention to use ICT among the owner of microenterprises were found to have a positive relationship. This is consistent with Khazaei (2020), who proclaimed that social influence could influence the enterprise's owner to use ICT. However, his study found that microenterprises have less and lower levels of ICT usage, while ICT users in medium and large enterprises have higher levels of ICT usage, and social factors influence the utilization of ICT among the enterprise. This explains that the management staffs and employees who are experts and have high profiles in technology would influence the intention of microenterprise owners to use ICT in their companies. Similarly, facilitating conditions of technology significantly affect the behavioral intention to use ICT. Razak et al. (2017) proclaimed that the current technical and organizational infrastructure can support the use of technology and influence enterprises to use ICT in their business operations. However, Kamaghe et al. (2020) proclaimed that a lack of guidance, timely support, information, and financial capacity might prohibit microenterprises from utilizing ICT systems in their companies.

From the result, one hypothesis was rejected: perceived ease of use. A possible explanation for why perceived ease of use does not influence behavioral intention of owners and managers of microenterprises includes despite knowing the ICT system is easy to learn. It would ease them to monitor day-to-day business operations, and the respondents proclaimed that they must frequently seek assistance when using ICT tools and systems. Although a previous study by Abdullah et al. (2021) suggested that perceived ease of use influences the behavioral intention among the SME owner to utilize ICT, this study found that ICT is not easy to use and tools are not friendly to users which may lead to refusal of using ICT among the microenterprises.

7 Conclusion

This study attempts to understand the behavioral intention of the owner and managers of microenterprises to use ICT in their companies. The factors of perceived usefulness, social influence, and facilitating condition should be strengthened to increase the usage of ICT among the owner of microenterprises in Malaysia. This is crucial to be considered by the government and ICT training providers to enhance the facilities' environment and support system of ICT to encourage more microenterprise owners to use ICT in their business operations. However, the researchers were challenged with limitations such as time. The research findings only represented students in the Faculty of Business and Entrepreneurship and Faculty of Hospitality, Tourism, and Wellness in one of Malaysia's public higher education institutions. Thus, it is recommended that students who own microenterprise in other public universities are included. Besides, future research is suggested to expand this study on other factors influencing behavioral intention to use ICT, such as areas of expertise, knowledge, attitude toward technology, and self-efficacy. By having this, the impact should be different from the current study.

References

Abdullah, H. H., Abdullah, S. S., Ramanchandram, R., & Ibrahim, M. (2021). The moderating influence of perceived usefulness in the adoption of technology by small and medium enterprises (SMEs): a comparative analysis between Malaysia and Nigeria. *Journal for Global Business Advancement, 14*(5), 705–728.

Ambarwati, R., Harja, Y. D., & Thamrin, S. (2020). The role of facilitating conditions and user habits: A case of Indonesian online learning platform. *Journal of Asian Finance, Economics and Business, 7*, 481–489. https://doi.org/10.13106/jafeb.2020.vol7.no10.481

Baig, F., & Shahzad, S. U. (2022). *Impact of social media influencer's credibility dimensions on consumer behavior: An empirical study related to influencer marketing on Pakistan's fashion industry*.

Basuki, R., Tarigan, Z. J., Siagian, H., Limanta, L. S., Setiawan, D., & Mochtar, J. (2022). The effects of perceived ease of use, usefulness, enjoyment and intention to use online platforms on behavioral intention in online movie watching during the pandemic era. *International Journal of Data and Network Science, 6*(1), 253–262. https://doi.org/10.5267/j.ijdns.2021.9.003

Buabeng-Andoh, C. (2018). Predicting students' intention to adopt mobile learning: A combination of theory of reasoned action and technology acceptance model. *Journal of Research in Innovative Teaching and Learning, 11*. https://doi.org/10.1108/JRIT-03-2017-0004

Busaidi, N. S., Bhuiyan, A. B., & Zulkifli, N. (2019). The critical review on the adoption of icts in the small and medium enterprises (smes) in the developing countries. *International Journal of Small and Medium Enterprises, 2*(2), 33–40. https://doi.org/10.46281/ijsmes.v2i2.437

Camilleri, M. A. (2020). The online users' perceptions toward electronic government services. *Journal of Information, Communication and Ethics in Society, 18*(2), 221–235.

Chan, F. K. Y., Thong, J. Y. L., Venkatesh, V., Brown, S. A., Hu, P. J. H., & Tam, K. Y. (2010). Modeling citizen satisfaction with mandatory adoption of an E-Government technology. *Journal of the Association for Information Systems, 11*(10), 519–549. https://doi.org/10.17705/1jais.00239

Damanpour, F., & Schneider, M. (2006). Phases of the adoption of innovation in organizations: Effects of environment, organization and top managers 1. *British Journal of Management, 17*(3), 215–236.

Davis, F. (1989). Perceived usefulness, perceived ease of use, and user acceptance of information technology. *MIS Quarterly, 13*, 319–340.

Dwivedi, Y., Rana, N., Jeyaraj, A., Clement, M., & Williams, and Michael. (2019). Re-examining the unified theory of acceptance and use of technology (UTAUT): Towards a revised theoretical model. *Information Systems Frontiers, 21*, 1–16. https://doi.org/10.1007/S10796-017-9774-Y

Gillies, F. I., Lye, C. T., & Tay, L. Y. (2020). Determinants of behavioral Intention to use bitcoin in Malaysia. *Journal of Information System and Technology Management, 5*(19), 25–38.

Gupta, S., et al. (2020). Dynamic capabilities and institutional theories for industry 4.0 and digital supply chain. *Supply Chain Forum: An International Journal, 21*(3), 139–157. https://doi.org/10.1080/16258312.2020.1757369

Handoko, B. L. (2019). Application of UTAUT theory in higher education online learning. In *Proceedings of the 2019 10th international conference on e-business, management and economics - ICEME 2019*. doi:https://doi.org/10.1145/3345035.3345047.

Hariri, A. A. (2014). *Adoption of learning innovations within UK Universities: Validating an extended and modified UTAUT mode* [Preprint]. doi: https://doi.org/10.1108/jtf-06-2021-0137

Huang, F., Teo, T., & Scherer, R. (2022). Investigating the antecedents of university students' perceived ease of using the Internet for learning. *Interactive Learning Environments, 30*, 1060–1076. https://doi.org/10.1080/10494820.2019.1710540

Kamaghe, J., Luhanga, E., & Kisangiri, M. (2020). The challenges of adopting M-learning assistive technologies for visually impaired learners in higher learning institution in Tanzania. *International Journal of Emerging Technologies in Learning, 15*(1), 140–151. https://doi.org/10.3991/ijet.v15i01.11453

Khazaei, H. (2020). Integrating cognitive antecedents to UTAUT model to explain adoption of blockchain technology among Malaysian SMEs. *JOIV: International Journal on Informatics Visualization, 4*(2), 85–90.

Kwabiah, W. A. (2019). *Small business strategies for information technology implementation in developing countries, ScholarWorks*. Available at: https://scholarworks.waldenu.edu/dissertations/7309

Masadeh, R., Tarhini, A., Mohammed, A. B., & Maqableh, M. (2016). Modeling factors affecting student's usage behaviour of E-learning systems in Lebanon. *International Journal of Business and Management, 11*(2), 299–312. https://doi.org/10.5539/ijbm.v11n2p299

Mwenda, M. J. (2018). *The use of information communication technology and performance of micro and small enterprises in Kenya: A case study of selected smes in Isiolo County, UoN Digital Repository Home*. University of Nairobi. Available at: http://erepository.uonbi.ac.ke/handle/11295/105481

Omar, F. I. (2018). Affective need of ICT in improving business performance among Malay women entrepreneur. *International Journal of Academic Research in Business and Social Sciences, 8*(6). https://doi.org/10.6007/ijarbss/v8-i6/4295

Orser, B., Riding, A., & Li, Y. (2019). Technology adoption and gender-inclusive entrepreneurship education and training. *International Journal of Gender and Entrepreneurship*. https://doi.org/10.1108/IJGE-02-2019-0026

Pedraza, J. M. (2021). The micro, small, and medium-sized enterprises and its role in the economic development of a country. *Business and Management Research, 10*(1), 33. https://doi.org/10.5430/bmr.v10n1p33

Pillay, P. (2016). *Barriers to information and communication technology (ICT) adoption and use amongst SMEs: A study of the South African manufacturing sector*. https://core.ac.uk/download/pdf/188775343.pdf.

Rahman, M. S., Das, S., Hossain, G. M. S., & Tajrin, T. (2021). Teenagers' behavioural intention towards wearable technologies and intention to recommend others: An empirical study in

Bangladesh. *Journal of Science and Technology Policy Management, 13*(1), 110–131. https://doi.org/10.1108/jstpm-05-2020-0088
Razak, F. Z. B. A., Bakar, A. A., & Abdullah, W. S. W. (2017). How perceived effort expectancy and social influence affects continuance of intention to use e-government. A study of a Malaysian government service. *Electronic Government, an International Journal, 13*(1). https://doi.org/10.1504/eg.2017.10004264
Rozmi, A. N., Bakar, M. I., Razak, A. Q., & Nordin, A. (2019). *Investigating the intentions to adopt ICT in Malaysian SMEs using the UTAUT Model.* IVIC. https://doi.org/10.1007/978-3-030-34032-2_42.
Sandberg, K. W., & Håkansson, F. (2020). Strategical use of ICT in microenterprises. *International Journal of E-Entrepreneurship and Innovation, 10*(1), 1–13. https://doi.org/10.4018/ijeei.2020010101
Subawa, N. S., Widhiasthini, N. W., & Mimaki, C. A. (2020, February). An empirical study of E-marketplace acceptance in MSMEs under the constructs of effort expectancy, social influence and facilitating condition factors. In *Proceedings of the 2020 The 6th international conference on E-business and applications* (pp. 116–120).
Suhartanto, D., & Leo, G. (2018). Small business entrepreneur resistance of ICT adoption: A lesson from Indonesia. *International Journal of Business and Globalisation, 21*, 5–18. https://doi.org/10.1504/ijbg.2018.10015253
Trivedi, S. K., & Yadav, M. (2020). Repurchase intentions in Y generation: Mediation of trust and e-satisfaction. *Marketing Intelligence and Planning, 38*, 401–415.
Tuncer, İ., Unusan, C., & Cobanoglu, C. (2021). Service quality, perceived value and customer satisfaction on behavioral intention in restaurants: An integrated structural model. *Journal of Quality Assurance in Hospitality and Tourism, 22*(4), 447–475. https://doi.org/10.1080/1528008X.2020.1802390
Venkatesh, V., Morris, M. G., Davis, G. B., & Davis, F. D. (2003). User acceptance of information technology: Toward a unified view. *MIS Quarterly*, 425–478.
Venkatesh, V., Thong, J. Y. L., & Xu, X. (2016). Unified theory of acceptance and use of technology: A synthesis and the road ahead. *Journal of the Association for Information Systems, 17*(5), 328–376. https://www.proquest.com/scholarly-journals/unified-theory-acceptance-use-technology/docview/1794948207/se-2
Yueh, H.-P., Huang, J.-Y., & Chang, C. (2015). Exploring factors affecting students' continued wiki use for individual and collaborative learning: An extended utaut perspective. *Australasian Journal of Educational Technology, 31*(1). https://doi.org/10.14742/ajet.170
Yunis, M., Tarhini, A., & Kassar, A. (2018). The role of ICT and innovation in enhancing organizational performance: The catalysing effect of corporate entrepreneurship. *Journal of Business Research, 88*, 344–356. https://doi.org/10.1016/j.jbusres.2017.12.030
Zuniarti, I., Yuniasih, I., Martana, I. K., Setyaningsih, E. D., Susilowati, I. H., Pramularso, E. Y., & Astuti, D. (2021). The effect of the presence of e-commerce on consumer purchasing decisions. *International Journal of Data and Network Science*, 479–484. https://doi.org/10.5267/j.ijdns.2021.3.005

Tahirah Abdullah is a senior lecturer at the Faculty of Entrepreneurship and Business, Universiti Malaysia Kelantan (UMK). She holds a Bachelor's degree in Accounting from Universiti Sains Islam Malaysia (USIM) and a Master's in Accounting from Universiti Islam Antarabangsa Malaysia (UIAM). She received her Doctor of Philosophy (PhD) in Accounting from Aberdeen, Scotland. She has been involved in many research projects and released journal articles, chapters in book and conference journals nationally and internationally, the Scopus and WOS index. Among her areas of expertise are capital management, capital management among Micro, Small, and Medium Enterprises (MSMEs), accounting in neoliberal enterprises, entrepreneurship, and education in accounting.

Siti Afiqah Zainuddin is a Senior Lecturer at the Faculty of Entrepreneurship and Business at Universiti Malaysia Kelantan (UMK). With over a decade of experience in teaching, she instructs students enrolled in the Bachelor of Accounting and Master of Entrepreneurship programs. Her academic qualifications include a Bachelor's Degree in Accounting (Honors) from the Islamic Science University of Malaysia, a Master's Degree in Science of Accounting from the International Islamic University of Malaysia, and a PhD in Accounting from the University of Aberdeen, Scotland, United Kingdom. She has made noteworthy contributions to SCOPUS and Web of Science-indexed publications. Additionally, she has been awarded more than 15 research grants for various accounting and entrepreneurship projects.

Amira Jamil is currently serving as a Senior Lecturer at the Faculty of Entrepreneurship and Business, Universiti Malaysia Kelantan (UMK). She earned her Doctor of Philosophy from Universiti Sultan Zainal Abidin in 2021, following a Master of Science (Accounting) in 2016 and a Bachelor of Accounting (Honors) in 2013, both from the International Islamic University Malaysia. She has published numerous academic papers and books and actively participated in various international conferences around the globe. Her areas of expertise encompass accounting, Islamic accounting, reporting, corporate governance, and taxation, with a keen interest in management, digital entrepreneurship, and digital marketing.

Siti Salwani Abdullah is a Senior Lecturer at the University Malaysia Kelantan's Faculty of Entrepreneurship and Business (UMK). She holds a Doctor of Philosophy in Finance from UMK, a Master of Science in Finance from University Utara Malaysia, and a Bachelor of Business Administration with Honors in Finance from University Technology MARA. Her scholarly interests span a vast array of fascinating topics. She is especially interested in studying entrepreneurial finance in the context of alternative financing (crowdfunding) and financial technology (FinTech). Her commitment to these fields has been recognized, as she has been awarded more than six gold medals in national and international innovation competitions. These honors explicitly recognize her outstanding contributions to educational innovation. In addition to her remarkable research endeavors, she has made a significant impact on the academic community through her outstanding teaching skills. As a result of her excellence in both teaching and research, she has been entrusted with the responsibility of leading the Evaluation and Assessment Unit at the Centre for Academic Excellence and Development (CAED), UMK. This appointment reflects the university's high appreciation for her skills and dedication.

Nur Farahiah Azmi is a Senior Lecturer at the Faculty of Entrepreneurship & Business of Universiti Malaysia Kelantan. She is appointed Deputy Director of the Islamic Centre of Universiti Malaysia Kelantan. She got her Doctor of Philosophy (Finance) from Universiti Malaysia Kelantan. She also has a Bachelor of Accountancy and a Master of Science (Finance) from the International Islamic University Malaysia (IIUM). Her research interest is in Islamic Social Finance, especially waqf, and she is also interested in governance, non-profit organizations, and qualitative research.

Nur Izzati Mohamad Anuar is a distinguished senior lecturer at the Faculty of Entrepreneurship and Business at Universiti Malaysia Kelantan. Holding a Ph.D. in Management from the University of Glasgow, Scotland, her expertise lies in organizational learning and development. With a passion for advancing the field of entrepreneurship, she has contributed significantly through her research and publications, addressing critical aspects such as knowledge, learning, and technology processes. Her work has been published in reputable journals indexed by Scopus and Web of Science, showcasing her dedication to academic excellence.

Nik Alif Amri Nik Hashim is a Senior lecturer in the Department of Tourism Management, Faculty of Hospitality, Tourism and Wellbeing (FHPK), Universiti Malaysia Kelantan (UMK). He holds a Bachelor of Science in Tourism Management with a minor in Hospitality and a Master's Degree in Tourism Management from Universiti Teknologi MARA (UiTM). He has been involved in research projects and has released many articles nationally and internationally. His areas of expertise include Tourism and Hospitality Management, Tourism Technology, Culture and heritage, Travel Risk Perceptions, Destination Destinations, Tourist Travelers and Intimate Adventures, Event Management in Tourism, and Crisis and Disaster Management. He holds a Doctor of Philosophy (PhD) in Tourism & Hospitality at the University of Queensland, Australia.

Siti Rohana Mohamad is a senior lecturer in the business and accounting department, Faculty of Entrepreneurship and Business, Universiti Malaysia Kelantan (UMK). She obtained her Doctor of Philosophy from Universiti Sultan Zainal Abidin in 2023, following a Master of Science (Finance) from the International Islamic University Malaysia (IIUM) in 2013 and a Bachelor of Business Administration (Hons) (Finance) from Universiti Teknologi Mara (UiTM) in 2010. She has been involved in many research projects, presenting at conferences and publishing journal articles and books locally and abroad. Her research interests include Islamic finance, Waqf, Islamic philanthropy, social entrepreneurship, and giving behavior.

Digital Entrepreneurship in the Food Delivery Industry: Enhancing Customer Satisfaction Through Apps' Functional and Technical Quality

Amira Jamil, Siti Afiqah Zainuddin, Tahirah Abdullah, Siti Rohana Mohamad, Nur Farahiah Azmi, Siti Fariha Muhamad, and Siti Salwani Abdullah

Abstract This study investigated the impact of digital entrepreneurship in the food delivery industry on customer satisfaction through the functional and technical quality of food delivery mobile applications. Customer satisfaction is a critical component in quality management, and it is crucial to identify the needs and expectations of various customer segments. This study considered the quality of application functionality (ease of use, quality of information) and technical quality (quality of delivered services, quality of delivered food) as factors affecting customer satisfaction. A quantitative research design was employed, and primary data was collected through an online survey given randomly to 382 students of the UMK City Campus. IBM SPSS Statistics was used to analyze the data, and multiple regression analysis was conducted to test the hypotheses. The study findings suggest that the functional and technical quality of food delivery mobile application services significantly influences customer satisfaction. This study concludes that digital entrepreneurship has revolutionized the food delivery industry and presents new opportunities for entrepreneurs to innovate and succeed in the market. This study has implications for food and beverage entrepreneurs, food delivery service providers, riders, and researchers interested in exploring the impact of digital entrepreneurship in the food delivery industry on customer satisfaction.

A. Jamil · T. Abdullah · S. R. Mohamad · N. F. Azmi · S. F. Muhamad · S. S. Abdullah
Universiti Malaysia Kelantan, Pengkalan Chepa, Kota Bharu, Kelantan, Malaysia
e-mail: amira@umk.edu.my; tahirah@umk.edu.my; rohana@umk.edu.my; nurfarahiah@umk.edu.my; fariha@umk.edu.my; salwani.a@umk.edu.my

S. A. Zainuddin (✉)
Universiti Malaysia Kelantan, Pengkalan Chepa, Kota Bharu, Kelantan, Malaysia

Department Accounting, Economics and Finance, School of Business Administration, European University Cyprus, Engomi, Cyprus
e-mail: sitiafiqah@umk.edu.my

N. Hayat, N. C. Nawi (eds.), *Contemporary Issues in Entrepreneurship and Innovative Technology*, Contributions to Environmental Sciences & Innovative Business Technology, https://doi.org/10.1007/978-3-031-50927-8_8

Keywords Customer satisfaction · Technology · Apps' functional and technical quality · Food delivery · Digital entrepreneurship

1 Introduction

The food delivery industry has undergone a significant transformation with the advent of digital entrepreneurship. The online food delivery industry has created numerous opportunities for entrepreneurs, particularly those in micro and small businesses. According to a report by Statista, the global food delivery market is expected to reach $154 billion in 2023, a significant increase from $107 billion in 2019. This development can be attributed to rising convenience demand, particularly among younger consumers who prefer ordering food online rather than cooking at home or dining out. One of the main advantages of the online food delivery industry is that it has lowered the barriers to entry for small businesses. With the rise of food delivery platforms such as Grab Food and Foodpanda, small restaurants and food vendors can now reach a wider audience without investing in expensive marketing campaigns or brick-and-mortar locations. As a result, the number of small businesses in the food delivery industry has increased significantly, particularly in urban areas where food delivery services are in high demand.

Despite the growing demand for these services, there is a need to understand the impact of digital entrepreneurship on customer satisfaction through the functional and technical quality of food delivery mobile applications. Customer satisfaction can be defined as the value-added difference between a customer's expectation and experience; in other words, customers are satisfied when the product or service they consume meets or exceeds their expectations (Lee et al., 2022). Customer satisfaction with the application was positively correlated with their intention to use it in the future (Chittithaworn et al., 2019). Another study by Alsaghier and Hussain (2021) explored the impact of food delivery mobile application quality on customer loyalty. The study found that both functional and technical quality positively impacted customer loyalty, indicating that improving application quality can increase customer retention and long-term success for digital entrepreneurs.

Thus, understanding the impact of digital entrepreneurship on customer satisfaction through the functional and technical quality of food delivery mobile applications is critical for entrepreneurs operating in the digital economy. By improving application quality and meeting customers' needs and expectations, digital entrepreneurs can increase customer satisfaction and loyalty, leading to long-term success in the competitive food delivery industry.

The present paper starts with an introduction that provides an overview of the study. The problem statement is then presented, outlining the gap in existing research that this study aims to address. The research question and objectives are then presented, followed by the underpinning theory that guides the study. A thorough literature review of existing research is presented, followed by the research framework, which outlines the conceptual model and hypotheses of the study. The research design, including data collection methods, sampling design, and research instrument, is described. The measurement of variables is also discussed. In the findings section, data analysis is presented, followed by a discussion of the results. Finally, implications for the study are discussed, and a conclusion is drawn based on the findings.

1.1 Problem Statement

The food delivery service industry has become a significant aspect of people's lives, with doorstep delivery being the most highly ranked factor influencing consumers to use food ordering applications (Beliya et al., 2019). As the availability of technology-based platforms has increased, the efficiency of online food delivery applications has become a crucial aspect of customer satisfaction. With the rapid development of technological innovation, society has become more accessible, and the Internet has become one of the mediums that most people use along with technology. This has led to the rise of food delivery services, which have become a new trend in modern life.

The rise of online food marketing and ordering systems has created a new wave of digital entrepreneurship opportunities for individuals and businesses. With the availability of websites and mobile applications, users can now easily order food from their favorite restaurants or even local vendors, providing an excellent platform for these entrepreneurs to connect with potential customers (Kumar & Pansari, 2021). Moreover, this trend has also led to numerous job opportunities in the food delivery service sector. As per a recent report, the online food delivery industry is expected to create around three million jobs by 2023 (TMR Research, 2022).

The gap in research on digital entrepreneurship in the food delivery industry is the lack of studies that investigate the influence of functional and technical quality of food delivery apps on customer satisfaction among university students in Malaysia, particularly in the context of Universiti Malaysia Kelantan (UMK). However, numerous studies have been conducted to investigate the impact of food delivery apps on customer satisfaction (Beliya et al., 2019). These studies focused on the general population and did not specifically investigate the use of food delivery apps among university students. Furthermore, there is a scarcity of research on the service quality of the riders who deliver food, which is also an important factor affecting customer satisfaction in the food delivery industry (Rae Hooi, 2021). As a result, this research aims to bridge the gap by investigating the influence of food delivery apps' functional and technical quality and the service quality of riders on customer satisfaction among UMK students on the City Campus.

1.2 Research Question

The following research questions provide a more detailed outlook on the issues that are the focus of this study:

1. Does ease of use on the application have a relationship with customer satisfaction with food delivery service among UMK City Campus students?
2. Does information quality on the application have a relationship with customer satisfaction with food delivery service among UMK City Campus students?
3. Does the delivery service quality have a relationship with customer satisfaction with food delivery service among UMK City Campus students?
4. Does the quality of delivered food have a relationship with customer satisfaction with food delivery service among UMK City Campus students?

1.3 Research Objective

The aims of this study are as follows:

1. To determine the relationship between the application's ease of use and customer satisfaction with food delivery service among UMK City Campus students.
2. To determine the relationship between information quality on the application and customer satisfaction with food delivery service among UMK City Campus students.
3. To analyze the relationship between delivery service quality and customer satisfaction with food delivery service among UMK City Campus students.
4. To examine the relationship between the quality of delivered food and customer satisfaction with food delivery service among UMK City Campus students.

2 Literature Review

2.1 Underpinning Theory

The Unified Theory of Acceptance and Use of Technology (UTAUT) is a widely used theoretical framework that explains the factors influencing technology acceptance and use. In the context of this study, UTAUT can provide insights into how the functional and technical quality of food delivery mobile applications affect customer satisfaction. The UTAUT model proposes that four key constructs—performance expectancy, effort expectancy, social influence, and facilitating conditions—influence users' behavioral intentions and technology use (Venkatesh et al., 2003).

Performance expectancy refers to the perceived usefulness of a technology, which is affected by factors such as ease of use, quality of information, and system functionality (Venkatesh et al., 2003). In the context of mobile food delivery applications, customers expect the app to provide accurate and detailed information about the food, delivery time, and cost. The app's quality of information influences the user's perception of its usefulness, which affects their satisfaction with the service.

Effort expectancy refers to the perceived ease of use of a technology, which is influenced by complexity, clarity, and ease of learning (Venkatesh et al., 2003). In the context of food delivery mobile applications, customers expect the app to be easy to navigate and use. The app's interface should be simple, intuitive, and user-friendly, making it easy for customers to place orders and track their delivery status. Therefore, the app's ease of use affects the user's satisfaction with the service.

Social influence refers to the influence of social norms, subjective norms, and social support on users' behavioral intentions and technology use (Venkatesh et al., 2003). In the context of food delivery mobile applications, customers may be influenced by social norms, such as the opinions of their peers or family members, who may recommend certain apps based on their experience. Positive reviews and ratings on the app store may also influence the customer's decision to use the app. Social influence, therefore, can affect the user's behavioral intention to use the app and their satisfaction with the service.

Facilitating conditions refer to the external factors that affect the user's ability to use the technology, such as the availability of resources, infrastructure, and technical support (Venkatesh et al., 2003). In the context of food delivery mobile applications, the quality of delivered services and the quality of delivered food are external factors that can influence the user's satisfaction with the service. For instance, if the app promises fast delivery, but the food arrives cold or late, the user's satisfaction may be affected.

In conclusion, the UTAUT model provides a valuable framework for understanding how food delivery mobile applications' functional and technical quality affect customer satisfaction. By considering the factors that influence performance expectancy, effort expectancy, social influence, and facilitating conditions, food and beverage entrepreneurs, food delivery service providers, and researchers can develop strategies to enhance customer satisfaction and improve the overall quality of the service.

A vast body of literature has used the Unified Theory of Acceptance and Use of Technology (UTAUT) framework to investigate the relationship between technology acceptance and customer satisfaction. In the food delivery industry, a study by Han et al. (2020) used UTAUT to examine the factors influencing the adoption of food delivery mobile applications in South Korea. They found that performance expectancy, effort expectancy, and social influence were significant predictors of behavioral intention, which in turn influenced customer satisfaction with the service. This study suggests that food delivery service providers should focus on improving the quality of information and ease of use of their mobile applications to enhance customer satisfaction.

Another study by Luo et al. (2021) used UTAUT to investigate the factors influencing the adoption of food delivery mobile applications in China. They found that performance expectancy, effort expectancy, and facilitating conditions were significant predictors of behavioral intention, which in turn influenced customer satisfaction with the service. This study highlights the importance of considering external factors such as delivery time and food quality in enhancing customer satisfaction.

In conclusion, prior literature has shown that UTAUT is a valuable framework for understanding the factors influencing technology acceptance and customer satisfaction in various industries, including food delivery. By considering the factors that influence performance expectancy, effort expectancy, social influence, and facilitating conditions, service providers can develop strategies to enhance customer satisfaction and improve the overall quality of the service.

2.2 *Previous Study*

Digital entrepreneurship has revolutionized the food delivery industry, providing new opportunities for entrepreneurs to innovate and succeed in the market. With the rise of mobile technology, food delivery applications have emerged, changing how food is delivered to customers. According to Technomic (2019), the food delivery market in the United States is projected to reach $76 billion by 2022, creating a promising landscape for digital entrepreneurs looking to tap into this market.

The rise of digital entrepreneurship has enabled food delivery companies to expand their reach and target customers through mobile applications that offer a range of features such as real-time tracking of orders, personalized menus, and customer feedback. According to a study by McKinsey and Company, customers who use food delivery applications are more likely to order frequently, spend more per order, and are less likely to abandon their orders (McKinsey & Company, 2018). Therefore, providing high-quality mobile application services is crucial for food delivery companies to enhance customer satisfaction and increase their market share.

Customer Satisfaction Customer satisfaction in the food delivery industry refers to the degree to which customers' needs and expectations are met through the delivery of food and related services. According to Murfield et al. (2017), satisfaction is often seen as an overall evaluation of an organization's product or service performance. Entrepreneurs in the food delivery industry need to understand customers' needs and expectations to provide high-quality services and build a loyal customer base. A satisfied customer is more likely to become a repeat customer, recommend the service to others, and leave positive reviews, contributing to the growth of the business.

The review of prior literature suggests that various factors influence customer satisfaction in the food delivery industry. Nguyen et al. (2018) found that tangibles (physical attributes of service quality) were the most important factor driving customer satisfaction in the UK fast-food industry. Katoch and Sidhu (2021) found that in India's online food delivery industry, product quality and pricing were the most critical factors for customer satisfaction, followed by consumer convenience. Overall, the papers suggest that various factors, including food quality, pricing, convenience, and service quality, influence customer satisfaction in the food delivery industry. Nevertheless, there is a significant lack of literature on the relationship between food delivery apps' functional and technical quality and customer satisfaction.

Apps' Functional Quality Growing research examines the relationship between app functional quality, precisely ease of use and information quality, and customer satisfaction with food delivery services. Ease of use refers to the degree to which an app is user-friendly, intuitive, and easy to navigate, while information quality refers to the accuracy and relevance of the information provided to users.

One study by Gao et al. (2020) found that ease of use significantly and positively influenced customer satisfaction with food delivery apps. The study also highlighted the importance of app performance, suggesting that fast and responsive apps are more likely to satisfy customers. Similarly, another study by Li and Liang (2018) found that perceived quality, including ease of use and information quality, significantly influenced customer satisfaction.

A study by Chen et al. (2020) explored the impact of information quality on customer satisfaction with food delivery apps. The study found that information quality significantly influenced customer satisfaction, with accurate and relevant information particularly important. Moreover, a study by Xia et al. (2021) examined the impact of app features, including ease of use and information quality, on customer satisfaction with food delivery apps. The study found that ease of use and personalized recommendations significantly influenced customer satisfaction, while information quality did not significantly impact satisfaction.

Overall, research suggests that ease of use and information quality contribute to customer satisfaction with food delivery apps. Apps that are user-friendly and responsive and provide accurate and relevant information are more likely to satisfy customers, ultimately increasing customer loyalty and retention. Therefore, these two hypotheses are developed:

H1: There is a positive correlation between customer satisfaction and the ease of use of food delivery service applications among UMK City Campus students.

H2: There is a positive correlation between customer satisfaction and the quality of the information provided for delivery services in using food delivery service applications among UMK City Campus students.

Technical Quality Food delivery service has become an integral part of the food industry and has rapidly grown in popularity. The technical quality of food delivery service, which encompasses the quality of the delivery service and the quality of the delivered food, is critical to customer satisfaction. Delivery service quality refers to the level of service provided by the delivery staff, including their attitude, reliability, and timeliness. On the other hand, the quality of the delivered food refers to the food's taste, freshness, and presentation. Both aspects are critical to the overall quality of the food delivery service.

According to a study by Kim and Lee (2021), food delivery service quality significantly impacts customer satisfaction. The study found that delivery service and food quality positively affect customer satisfaction. The study also found that delivery service quality had a more significant impact on customer satisfaction than delivered food quality. This suggests that while the quality of the delivered food is essential, customers also value reliable and timely delivery service. Another study by Anwar et al. (2021) found that the reliability of delivery service significantly influenced customer satisfaction. The study also found that the quality of the delivered food significantly impacted customer satisfaction.

In contrast to Kim and Lee's (Kim & Lee, 2021) study, Anwar et al. (2021) found that the quality of the delivered food had a more significant impact on customer satisfaction than the delivery service quality. Furthermore, a study by Lee et al. (2021) found that the presentation of the delivered food significantly influenced customer satisfaction. The study found that customers value food that is well-presented and visually appealing. This suggests that the quality of the delivered food is not just about taste and freshness but also about presentation.

In conclusion, the technical quality of food delivery service, including the quality of the delivery service and the quality of the delivered food, significantly impacts customer satisfaction. While different studies have found varying degrees of influence for each aspect, it is clear that both delivery service quality and delivered food quality are essential factors in determining customer satisfaction. Therefore, these two hypotheses are developed:

H3: There is a positive correlation between customer satisfaction and the delivery service quality in using food delivery service applications among UMK City Campus students.

H4: There is a positive correlation between customer satisfaction and the quality of delivered food using food delivery service applications among UMK City Campus students.

2.3 Research Framework

A comprehensive research framework has been developed to understand the relationship between independent and dependent variables. The dependent variable of interest in this model is customer satisfaction. In contrast, the independent variables include app functional quality, further subdivided into ease of use and information quality, and technical quality, further subdivided into the quality of delivered service and the quality of delivered food. Based on previous research findings, this conceptual model can help shed light on the relationships between these variables and provide insights into the factors contributing to customer satisfaction in the food delivery industry. By examining how app functional quality and technical

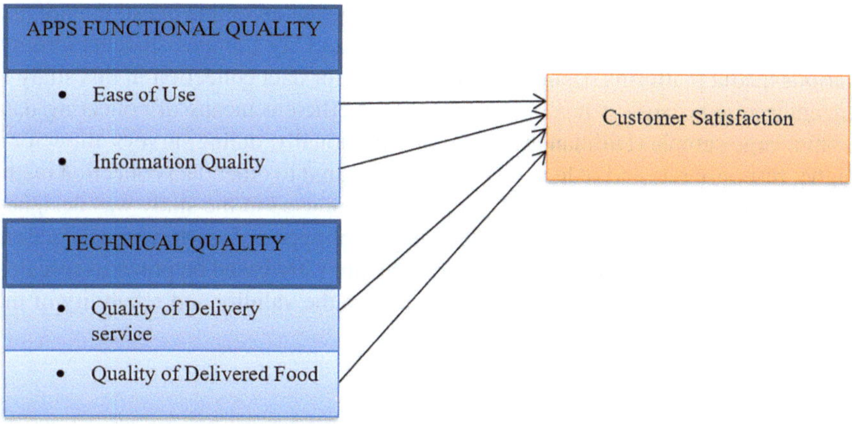

Fig. 1 Research framework

quality can impact customer satisfaction, this model can help to identify areas for improvement and inform strategies for enhancing customer experiences, thus improving the practice of digitalized entrepreneurship. The research framework for this investigation is displayed in Fig. 1.

3 Research Methodology

3.1 Research Design

Research design is the systematic and structured approach to planning, executing, and reporting a research study. It refers to the procedures, methods, and techniques used to collect, analyze, and interpret data to answer the research question or hypothesis (Creswell & Creswell, 2017). Since the primary purpose of this study is to identify levels and the nature of certain relationships, particularly between app functional and technical quality with customer satisfaction with food delivery service, the researchers choose descriptive research as the appropriate design of this study. This design is appropriate for exploring the relationship between variables, such as app functional and technical quality and customer satisfaction.

3.2 Data Collection

Data collection is a crucial initial stage of research that can significantly impact the usefulness of findings by minimizing the likelihood of errors during the investigation (Sekaran & Bougie, 2016). In this study, the researchers focused on primary data collection. An online survey was conducted using an online questionnaire distributed randomly to University Malaysia Kelantan City Campus students through various media platforms. Using an online survey for data collection in this study is advantageous because it is a cost-effective and efficient means of collecting data from a large sample (Dillman et al., 2014). Additionally, online surveys allow for a higher response rate and reduce survey bias compared to other data collection methods (Krosnick & Presser, 2010). The questionnaire used in the study was designed to measure the functional and technical quality of apps and customer satisfaction. The questions included in the survey questionnaire were based on previous research on the topic and were carefully crafted to ensure the validity and reliability of the data collected (Hair et al., 2019).

3.3 Sampling Method

This study's population is all UMK students on the City Campus. Based on the Academic Administration Division, the total population was 6013 students, as illustrated in Table 1.

According to Krejcie and Morgan (1970), the sample size (S) should be determined based on the population size (N). This study's population is 6013 UMK City Campus students; thus, the sample size will be 361. Convenience sampling was chosen as the sampling method in this study due to the practical limitations of time, cost, and the large sample size of the population. According to Cohen et al. (2018), convenience sampling is commonly used when the population size is large, and it is not feasible to select a random sample due to the time and cost constraints of the research.

3.4 Procedure for Data Analysis

The data collected from the online questionnaire in this study were analyzed using the Statistical Package for the Social Sciences (SPSS) software, which is widely used in social science research. Data analysis aimed to explore the study findings and answer the research question.

The analysis phase of the study involved several statistical techniques, including descriptive statistics, the Pearson correlation coefficient, and multiple linear

Table 1 Total population of faculties in UMK City Campus

Faculty	Population
Business school	3495
Hospitality school	2518
Total population	6013

regression. Descriptive statistics were used to summarize the characteristics of the data, such as measures of central tendency and variability. The Pearson correlation coefficient examined the association between two nonparametric variables, and multiple linear regression was used to examine the relationship between the dependent variable and multiple independent variables.

4 Findings

4.1 Demographic Profile of the Respondents

Table 2 illustrates the demographic profile of the respondents. It can be concluded that there are more female respondents (51.8%) than male respondents (48.2%). In terms of age, the highest percentage is 21–23 years which is 71.7%, followed by 24 years and above (14.7%) and 18 to 20 years (13.6%). The majority of the respondents were Malay (76.4%), followed by Chinese (11.5%), Indian (9.4%), and others (2.6%). Regarding marital status, there are two: single (94.5%) and married (5.5%).

The course of study in this study is divided into six different courses. The highest percentage is from SAR, which is 38.0%, followed by SAB (18.6%), SAK (14.1%), SAL (13.9%), SAE (8.6%), and SAA (6.8%). In terms of previous education level, the majority of respondents are from STPM, which is 54.7%, followed by Matriculation (20.2%), Diploma (10.2%), Foundation (8.1%), and STAM (6.8%). Next, about income, the highest percentage is from study-loan, which is 60.0%, followed by scholarship (36.4%), family/parents (1.8%), self-sponsored (1%), and none (0.8). We also can see that the most preferred food delivery service and the highest come from Foodpanda, which is 55.0%, followed by Grab Food (44.8%) and ShopeeFood (0.3%). Regarding the frequency of using food delivery service, the highest percentage is rarely (less than one time per month) which is 40.3% followed by more than two times per week (24.1%), one to two times per month (18.3%), and one to two times per week (17.3%).

4.2 Reliability Test

Reliability is a critical aspect of any research study, and it refers to the consistency and stability of research instruments, such as tests and scales used in data collection (Carmines & Zeller, 1979). One commonly used tool for assessing the reliability of research instruments is Cronbach's alpha coefficient. Cronbach's alpha coefficient measures the internal consistency of items intended to measure a particular construct or variable (Tavakol & Dennick, 2011). In other words, it measures how closely related a group of items are and whether they measure the same underlying construct. According to Tavakol and Dennick (2011), Cronbach's alpha coefficient

Table 2 Demographic profile of the respondents

		N	%
Gender	Male	184	48.2
	Female	198	51.8
Age	18–20 years	52	13.6
	21–23 years	274	71.7
	24 years and above	56	14.7
Marital status	Single	361	94.5
	Married	21	5.5
Course attend	SAB	71	18.6
	SAL	53	13.9
	SAR	145	38.0
	SAK	54	14.1
	SAE	33	8.6
	SAA	26	6.8
Previous education level	Matriculation	77	20.2
	Foundation	31	8.1
	STAM	26	6.8
	STPM	209	54.7
	Diploma	39	10.2
Income	Scholarship	139	36.4
	Study-loan	229	60.0
	Family/parents	7	1.8
	Self-sponsored	4	1.0
	None	3	0.8
Most preferred food delivery service	Grab Food	171	44.8
	Foodpanda	210	55.0
	Others	1	0.3
Frequency of using food delivery service	More than two times per week	92	24.1
	One to two times per week	70	18.3
	One to two times per month	66	17.3
	Rarely (less than one time per month)	154	40.3

can range from 0 to 1, with values closer to 1 indicating greater internal consistency. A Cronbach's alpha coefficient of 0.7 or above is generally acceptable for research purposes (Nunnally, 1978).

Table 3 shows the result of the reliability analysis. The indicated Cronbach's alpha value for customer satisfaction of food delivery service among UMK Campus Kota students, ease of use, information quality, and delivery service quality is more than 0.7 and is considered reliable.

4.3 Pearson Correlation Coefficient

Pearson correlation is one of the measures used to measure the strength of a linear relationship between two variables. The study used this analysis to identify the significant relationship between the dependent variable (customer satisfaction) and independent variables (ease of use, information quality, quality of delivered service, and quality of delivered food).

Based on the result illustrated in Table 4, it can be concluded that both apps' functional and technical quality has a significantly positive relationship with customer satisfaction. The relationship between apps' functional quality, represented by the ease of use and information quality, has a greater degree of significance with customer satisfaction, as portrayed by the significance value of 0.785 ($p < 0.01$).

4.4 Multiple Linear Regression Analysis

Multiple linear regression is a statistical technique to explore the relationship between a single dependent variable and two or more independent variables. In this study, multiple linear regression is used to predict customer satisfaction based on four independent variables: ease of use, information quality, quality of delivered service, and quality of delivered food among UMK City Campus students.

From Table 5, the result shows that the *P* value of the ease of use and information quality is 0.000, where the value is less than α (0.05). It indicates that ease of use and information quality have a significant positive relationship with customer satisfaction in using the mobile applications of food delivery services. The results are consistent with the initial Pearson correlation analysis.

While the *P* values of the quality of delivered service and quality of delivered food are 0.016 and 0.002, respectively, it is explained that these variables are also significant factors toward the food delivery service. Based on the observation, ease of use and information quality on the application are more significant than other variables, and it also has a significant positive relationship with customer satisfaction. This is because the value is 0.000, less than 0.05. The overall result of the hypothesis testing is presented in Table 6.

Table 3 Summary of reliability test

Variables	Cronbach's alpha	Number of items	Strength
Customer satisfaction	0.921	6	Excellent
Ease of use	0.901	5	Excellent
Information quality	0.909	5	Excellent
Quality of delivered service	0.903	5	Excellent
Quality of delivered food	0.910	5	Excellent

Table 4 The result of Pearson correlation coefficient

		Customer satisfaction	Ease of use	Information quality	Quality of delivered service	Quality of delivered food
Customer satisfaction	Pearson correlation	1	0.785**	0.785**	0.737**	0.734**
	sig. (two-tailed)		0.000	0.000	0.000	0.000

Table 5 The result of multiple linear regression analysis

	Unstandardized coefficient		Standardized coefficients		
	B	Std. error	*B*	*t*	Sig.
(Constant)	0.198	0.143		1.382	0.168
Ease of use	0.350	0.065	0.319	5.365	0.000
Information quality	0.280	0.065	0.268	4.302	0.000
Quality of delivered service	0.143	0.059	0.139	2.427	0.016
Quality of delivered food	0.178	0.058	0.173	3.082	0.002

5 Discussion and Conclusion

The primary objective of this study is to investigate the relationship between customer satisfaction and the apps' functional quality and technical quality of food delivery services. Four independent variables were tested: ease of use, information quality, quality of delivered food, and quality of delivered services. Primary data was collected via an online questionnaire, and the findings were analyzed using SPSS. According to the Pearson correlation and multiple regression analyses, all variables had a significant positive relationship with customer satisfaction with food delivery services via mobile applications.

The significant relationship between the ease of use of food delivery apps and customer satisfaction suggests that app developers and food delivery companies should prioritize the user experience when designing their apps. The findings of this study align with previous research that has identified ease of use as a crucial factor in shaping customer satisfaction in various industries, including the food delivery industry (Alalwan et al., 2017; Chen & Dubinsky, 2003). In today's fast-paced digital age, where customers have high expectations for the functionality of their apps, businesses must prioritize ease of use to remain competitive. Doing so can attract and retain customers by offering them an intuitive, user-friendly experience that simplifies ordering food delivery services. Additionally, a user-friendly app may help to increase customer loyalty and repeat business, leading to better business outcomes for food delivery companies.

Additionally, the study found a significant relationship between information quality and customer satisfaction. This finding aligns with previous research highlighting the importance of accurate and reliable information in enhancing customer satisfaction (Paraskevas & Arendell, 2007). Customers rely on the information

Table 6 The result of the hypothesis testing

H	Statement of hypothesis	Results
H1	There is a positive relationship between ease of use and customer satisfaction with food delivery service among UMK City Campus students	Supported
H2	There is a positive relationship between information quality and customer satisfaction with food delivery service among UMK City Campus students	Supported
H3	A positive relationship exists between the quality of delivered food and customer satisfaction with food delivery service among UMK City Campus students	Supported
H4	There is a positive relationship between the quality of food delivery and customer satisfaction with food delivery service among UMK City Campus students	Supported

provided by the app to make informed decisions about their food orders, such as menu options, nutritional information, and delivery times. When the information provided is inaccurate or incomplete, customers may become dissatisfied with the service and choose not to use the app in the future. This study's findings suggest that food delivery companies should prioritize providing high-quality and reliable information through their apps to enhance customer satisfaction. By ensuring the information is accurate, up-to-date, and easily accessible, companies can build trust with their customers and establish a positive reputation for their brand. Providing clear and detailed information can also help to reduce confusion and errors, leading to a smoother ordering and delivery process.

Furthermore, the study identified a significant relationship between delivery service quality and customer satisfaction. This finding supports previous studies that have emphasized the role of delivery service quality in customer satisfaction (Baumann et al., 2013). The present study reinforces the notion that food delivery companies must prioritize delivery service quality to ensure customer satisfaction. Companies can achieve this by ensuring timely and accurate delivery, ensuring delivery personnel are polite and professional, and allowing customers to track their orders. Failure to prioritize delivery service quality can lead to dissatisfied customers and negative online reviews, harming a company's reputation and ultimately impacting its bottom line.

Lastly, the study found a significant relationship between the quality of delivered food and customer satisfaction. This finding is consistent with previous research on the importance of food quality in determining customer satisfaction in the food industry (Baumann et al., 2013; Kim & Jang, 2018). Customers expect the food they receive to be of high quality, and if their expectations are met, they are more likely to be satisfied with the service. This highlights the need for food delivery companies to prioritize the quality of their food products to improve customer satisfaction.

Overall, the results of this study suggest that enhancing the functional and technical quality of food delivery apps can lead to increased customer satisfaction in the food delivery industry. This has important implications for digital entrepreneurship in the food delivery industry, as it highlights the importance of investing in app development and delivery service quality to enhance customer satisfaction and gain a competitive advantage.

6 Implications of the Study

The study has several implications for food and beverage entrepreneurs, food delivery service providers, riders, and researchers interested in the impact of digital entrepreneurship on customer satisfaction in the food delivery industry.

Firstly, the study highlights the importance of the functional and technical quality of food delivery mobile application services in enhancing customer satisfaction. Entrepreneurs and service providers must focus on providing easy-to-use apps with accurate and relevant information on delivery services and food quality to ensure customer satisfaction. The study also emphasizes the role of digital entrepreneurship in revolutionizing the food delivery industry. Entrepreneurs leveraging technology to improve service quality and differentiate themselves from competitors can succeed in the market.

The study provides insight into customer needs and expectations in the food delivery industry. Entrepreneurs and service providers can use this information to tailor their services to specific customer segments and increase customer satisfaction. It also highlights the importance of quality management in the food delivery industry. To maintain customer satisfaction and loyalty, entrepreneurs and service providers must prioritize quality in all aspects of their service, including app functionality, delivery services, and food quality.

Finally, the study contributes to the growing body of research on the impact of digital entrepreneurship on customer satisfaction in the food delivery industry. Further research can build on these findings and explore other factors influencing customer satisfaction in this dynamic and rapidly evolving industry.

7 Recommendation

Based on the findings of this study, several recommendations for future research could extend and improve upon the present study by investigating the role of personalization in enhancing customer satisfaction in food delivery apps. Personalization is crucial to customer satisfaction in various industries (Lemon & Verhoef, 2016), and it could be particularly relevant in the food delivery industry, where customers have unique preferences and dietary restrictions.

Besides, future research could examine the impact of app design on customer satisfaction. App design, including visual esthetics, layout, and navigation, has influenced user satisfaction and usage (Huang et al., 2018). Future studies could explore how app design could be optimized to enhance customer satisfaction in the food delivery industry.

Future research could also conduct a comparative analysis of customer satisfaction across different food delivery apps. This could involve comparing customer satisfaction across apps with different business models (e.g., commission-based vs. subscription-based), delivery methods (e.g., in-house delivery vs. third-party delivery), and user interfaces. Such studies could provide insights into factors contributing to customer satisfaction in different contexts and inform app development and marketing strategies.

Moreover, future research could investigate the impact of digital entrepreneurship on other aspects of the food delivery industry, such as sustainability and social responsibility. The food delivery industry has been criticized for its environmental impact and labor practices (Deutz et al., 2021). Future studies could explore how digital entrepreneurship could be leveraged to address these issues and promote sustainable and socially responsible practices in the industry.

Acknowledgments The authors gratefully acknowledge the financial support from the Universiti Malaysia Kelantan under the UMK Fundamental (UMK-FUND) Grant Scheme: R/FUND/A0100/01049A/001/2022/01105.

References

Alalwan, A. A., Dwivedi, Y. K., & Rana, N. P. (2017). Factors influencing adoption of mobile banking by Jordanian bank customers: Extending UTAUT2 with trust. *International Journal of Information Management, 37*(3), 99–110.

Alsaghier, H. A., & Hussain, S. (2021). Impact of mobile application quality on customer loyalty: A study of food delivery apps in Saudi Arabia. *International Journal of Information Management, 57*, 102289.

Anwar, S., Mehmood, S., & Hussain, S. (2021). Service quality dimensions and customer satisfaction in the food delivery industry. *Journal of Foodservice Business Research, 24*(2), 203–220. https://doi.org/10.1080/15378020.2020.1857151

Baumann, C., Burton, S., & Elliott, G. (2013). Modelling customer satisfaction and loyalty: Survey data versus data-mining. *Journal of Services Marketing, 27*(4), 322–333.

Beliya, A., Kujur, R., Verma, M., Nagwanshi, K. V., Sahu, S., Uikey, N., & Bhat, A. A. (2019). Satisfaction of consumers by using online food services. *International Journal of Humanities and Social Sciences, 8*(4), 35–44.

Chen, Y., & Dubinsky, A. J. (2003). A conceptual model of perceived customer value in e-commerce: A preliminary investigation. *Psychology and Marketing, 20*(4), 323–347.

Chen, M., Wu, J., & Liang, H. (2020). The impact of information quality on customer satisfaction with food delivery apps. *Journal of Hospitality and Tourism Technology, 11*(3), 446–460.

Chittithaworn, C., Phusavat, K., & Khrueathai, P. (2019). Developing a food delivery app for enhancing customer satisfaction in Thailand. *Journal of Hospitality and Tourism Technology, 10*(3), 368–384.

Cohen, L., Manion, L., & Morrison, K. (2018). *Research methods in education*. Routledge.

Creswell, J. W., & Creswell, J. D. (2017). *Research design: Qualitative, quantitative, and mixed methods approaches*. Sage.

Deutz, P., Mies, R., & Offenberger, U. (2021). The environmental impact of food delivery: An analysis of carbon emissions and labor practices. *Journal of Cleaner Production, 279*, 123606.

Dillman, D. A., Smyth, J. D., & Christian, L. M. (2014). *Internet, phone, mail, and mixed-mode surveys: The tailored design method* (4th ed.). Wiley.

Gao, X., Chen, S., & Wan, Y. (2020). Exploring the determinants of customer satisfaction with food delivery apps: An empirical study. *Journal of Retailing and Consumer Services, 52*, 101915.

Hair, J. F., Black, W. C., Babin, B. J., & Anderson, R. E. (2019). *Multivariate data analysis* (8th ed.). Cengage Learning.

Han, H., Lee, S., & Yoon, S. (2020). The effects of food delivery apps' perceived value, satisfaction, and quality on behavioral intention: A case of South Korea. *Sustainability, 12*(17), 6918.

Huang, Z., Benyoucef, M., & Liu, L. (2018). User satisfaction with mobile applications: A literature review and future directions. *International Journal of Information Management, 43*, 342–355.

Katoch, R., & Sidhu, A. (2021). *Online food delivery industry in India: A case of customer satisfaction dynamics.*

Kim, K. H., & Jang, S. S. (2018). Do food quality and service quality really matter in the restaurant industry? Moderating effects of restaurant type. *Journal of Quality Assurance in Hospitality and Tourism, 19*(1), 1–22.

Kim, H. J., & Lee, K. (2021). Determinants of customer satisfaction in the food delivery service industry: Focusing on delivery service quality and food quality. *Sustainability, 13*(3), 1283. https://doi.org/10.3390/su13031283

Krejcie, R. V., & Morgan, D. W. (1970). Determining sample size for research activities. *Educational and Psychological Measurement, 30*(3), 607–610. https://doi.org/10.1177/001316447003000308

Krosnick, J. A., & Presser, S. (2010). Question and questionnaire design. In *Handbook of survey research* (2nd ed., pp. 263–314). Emerald Group Publishing Limited.

Kumar, A., & Pansari, A. (2021). Digital entrepreneurship: Opportunities and challenges in the digital age. *Journal of Small Business and Entrepreneurship Development, 8*(1), 1–10.

Lee, J. H., Kim, J. H., & Lee, H. S. (2021). Analysis of customer satisfaction factors in the food delivery service industry: Focused on food quality and service quality. *Journal of Foodservice Management, 24*(2), 51–70. https://doi.org/10.21827/jfma.2021.24.2.51

Lee, V., Park, S., & Lee, D. (2022). The effect of E-commerce service quality factors on customer satisfaction, purchase intention, and actual purchase in Uzbekistan. *Global Business and Finance Review*, 56–74.

Lemon, K. N., & Verhoef, P. C. (2016). Understanding customer experience throughout the customer journey. *Journal of Marketing, 80*(6), 69–96.

Li, X., & Liang, X. (2018). The impact of perceived quality on customer satisfaction in the context of food delivery apps. *International Journal of Business and Management, 13*(10), 221–228.

Luo, M., Li, X., & Wu, Y. (2021). A study of the factors influencing the adoption of food delivery mobile applications in China. *Sustainability, 13*(10), 5741.

McKinsey & Company. (2018). *The state of global delivery*. Retrieved from https://www.mckinsey.com/~/media/McKinsey/Industries/Travel%20Transport%20and%20Logistics/Our%20Insights/The%20state%20of%20global%20delivery/The-state-of-global-delivery.ashx

Murfield, M., Boone, C. A., Rutner, P., & Thomas, R. (2017). Investigating logistics service quality in omni-channel retailing. *International Journal of Physical Distribution & Logistics Management, 47*(4), 263–296.

Nguyen, Q. H., Nisar, T. M., Knox, D., & Prabhakar, G. (2018). Understanding customer satisfaction in the UK quick service restaurant industry. *British Food Journal.*

Nunnally, J. C. (1978). *Psychometric theory* (2nd ed.). McGraw-Hill.

Paraskevas, A., & Arendell, B. (2007). Perceived website service quality in a tourism context. *The Service Industries Journal, 27*(6), 783–802.

Rae Hooi, T. K. (2021). Intention to use online food delivery services in Malaysia among university students. *Conference on management, business, innovation, education and social science.*

Sekaran, U., & Bougie, R. (2016). *Research methods for business: A skill building approach* (7th ed.). Wiley.

Tavakol, M., & Dennick, R. (2011). Making sense of Cronbach's alpha. *International Journal of Medical Education, 2*, 53–55. https://doi.org/10.5116/ijme.4dfb.8dfd

Technomic. (2019). *Food delivery in the US: A review of the restaurant delivery market and related trends*. Retrieved from https://www.technomic.com/Reports_and_Newsletters/Industry_Reports/dyn_PubLoad.php?pID=771

TMR Research. (2022). *Online food delivery market to generate 3 million jobs by 2023, states report*. Retrieved from https://www.tmrblog.com/2022/01/online-food-delivery-market-to-generate.html

Venkatesh, V., Morris, M. G., Davis, G. B., & Davis, F. D. (2003). User acceptance of information technology: Toward a unified view. *MIS Quarterly*, 425–478.

Xia, L., Liu, Y., & Guo, Y. (2021). The impact of app features on customer satisfaction in food delivery apps. *Journal of Retailing and Consumer Services, 61*, 102577.

Amira Jamil is currently serving as a Senior Lecturer at the Faculty of Entrepreneurship and Business, Universiti Malaysia Kelantan (UMK). She earned her Doctor of Philosophy from Universiti Sultan Zainal Abidin in 2021, following a Master of Science (Accounting) in 2016 and a Bachelor of Accounting (Honors) in 2013, both from the International Islamic University Malaysia. She has published numerous academic papers and books and actively participated in various international conferences around the globe. Her areas of expertise encompass accounting, Islamic accounting, reporting, corporate governance, and taxation, with a keen interest in management, digital entrepreneurship, and digital marketing.

Siti Afiqah Zainuddin is a Senior Lecturer at the Faculty of Entrepreneurship and Business at Universiti Malaysia Kelantan (UMK). With over a decade of experience in teaching, she instructs students enrolled in the Bachelor of Accounting and Master of Entrepreneurship programs. Her academic qualifications include a Bachelor's Degree in Accounting (Honors) from the Islamic Science University of Malaysia, a Master's Degree in Science of Accounting from the International Islamic University of Malaysia, and a PhD in Accounting from the University of Aberdeen, Scotland, United Kingdom. She has made noteworthy contributions to SCOPUS and Web of Science-indexed publications. Additionally, she has been awarded more than fifteen research grants for various accounting and entrepreneurship projects.

Tahirah Abdullah is a senior lecturer at the Faculty of Entrepreneurship and Business, Universiti Malaysia Kelantan (UMK). She holds a Bachelor's degree in Accounting from Universiti Sains Islam Malaysia (USIM) and a Master's in Accounting from Universiti Islam Antarabangsa Malaysia (UIAM). She received her Doctor of Philosophy (PhD) in Accounting from Aberdeen, Scotland. She has been involved in many research projects and released journal articles, chapters in book and conference journals nationally and internationally, the Scopus and WOS index. Among her areas of expertise are capital management, capital management among Micro, Small, and Medium Enterprises (MSMEs), accounting in neoliberal enterprises, entrepreneurship, and education in accounting.

Siti Rohana Mohamad is a senior lecturer in the business and accounting department, Faculty of Entrepreneurship and Business, Universiti Malaysia Kelantan (UMK). She obtained her Doctor of Philosophy from Universiti Sultan Zainal Abidin in 2023, following a Master of Science (Finance) from the International Islamic University Malaysia (IIUM) in 2013 and a Bachelor of Business Administration (Hons) (Finance) from Universiti Teknologi Mara (UiTM) in 2010. She has been involved in many research projects, presenting at conferences and publishing journal articles and books locally and abroad. Her research interests include Islamic finance, Waqf, Islamic philanthropy, social entrepreneurship, and giving behavior.

Nur Farahiah Azmi is a Senior Lecturer at the Faculty of Entrepreneurship & Business of Universiti Malaysia Kelantan. She is appointed Deputy Director of the Islamic Centre of Universiti Malaysia Kelantan. She got her Doctor of Philosophy (Finance) from Universiti Malaysia Kelantan. She also has a Bachelor of Accountancy and a Master of Science (Finance) from the International Islamic University Malaysia (IIUM). Her research interest is in Islamic Social Finance, especially waqf, and she is also interested in governance, non-profit organizations, and qualitative research.

Siti Salwani Abdullah is a Senior Lecturer at the University Malaysia Kelantan's Faculty of Entrepreneurship and Business (UMK). She holds a Doctor of Philosophy in Finance from UMK, a Master of Science in Finance from University Utara Malaysia, and a Bachelor of Business Administration with Honors in Finance from University Technology MARA. Her scholarly interests span a vast array of fascinating topics. She is especially interested in studying entrepreneurial finance in the context of alternative financing (crowdfunding) and financial technology (FinTech). Her commitment to these fields has been recognized, as she has been awarded more than six gold medals in national and international innovation competitions. These honors explicitly recognize her outstanding contributions to educational innovation. In addition to her remarkable research endeavors, she has made a significant impact on the academic community through her outstanding teaching skills. As a result of her excellence in both teaching and research, she has been entrusted with the responsibility of leading the Evaluation and Assessment Unit at the Centre for Academic Excellence and Development (CAED), UMK. This appointment reflects the university's high appreciation for her skills and dedication.

Leveraging Customer Relationship Management (CRM) for Stimulating Cyberpreneurship in Malaysia

Syamsuriana Sidek, Hazrina Hasbolah, Nur Syafiqah A. Samad, Zailani Abdullah, Nurul Husnina Najwa Zoraimi, Nurul Azwa Mohamed Khadri, Muhamad Ali Imran Kamarudin, and Nor Hizami Hassin

Abstract Customer relationship management (CRM) is frequently referred to as technological systems. Companies assist in griping external customer engagement at all customer life cycle stages, from discovery to education, purchase, and post-purchase. Social media is one approach of CRM that companies can adopt to trends that benefit their bottom line. However, the lack of current studies concerns CRM research. This study determines the roles of CRM in stimulating cyberpreneurship in Malaysia by determining the nexus between pricing strategy, quality assurance, customer engagement, purchase intention, and CRM. Quantitative research methods were used to collect empirical data from 340 selected online shoppers through purposive sampling techniques. Data were collected through questionnaires and initially analyzed descriptive statistics, validity, reliability, and normality test using IBM SPSS Statistics. The study hypotheses were analyzed using multiple linear regression, which confirmed that pricing strategy, assurance, customer engagement, and purchase intention directly affect customer relationship management. By examining the relationship between CRM and these digital marketing factors, the study provides insights into how cyberpreneurs can leverage CRM tools and strategies to enhance their online business activities. The findings can help cyberpreneurs understand the importance of implementing CRM systems, optimizing sales and marketing efforts, and improving customer retention in digital entrepreneurship. Therefore, this study contributes to understanding how CRM can support and drive cyberpreneurship endeavors in the digital marketplace

S. Sidek (✉) · H. Hasbolah · N. S. A. Samad · Z. Abdullah · N. H. N. Zoraimi · N. A. M. Khadri · N. H. Hassin
Universiti Malaysia Kelantan, Pengkalan Chepa, Kota Bharu, Kelantan, Malaysia
e-mail: suriana@umk.edu.my

M. A. I. Kamarudin
Universiti Utara Malaysia, Sintok, Kedah, Malaysia

N. Hayat, N. C. Nawi (eds.), *Contemporary Issues in Entrepreneurship and Innovative Technology*, Contributions to Environmental Sciences & Innovative Business Technology, https://doi.org/10.1007/978-3-031-50927-8_9

Keywords Cyberpreneurship · Customer relationship management · Pricing strategy · Quality assurance · Customer engagement · Purchase intention · Digital marketing · Information system

1 Introduction

Since the last decades, digital technologies such as electrical gadgets, devices, systems, or any resources that generate, store, or process data are becoming necessary tools as assistants to human life. Not only limited to making human work easier, digital technologies, including cyberpreneurship activities, are crucial to improving business organization performance. As proof, 90% of enterprises implement digital strategies to achieve a competitive advantage and sustain in the market (Świeczak & Łukowski, 2016).

Cyberpreneurship in Malaysia has gained significant attention in recent years due to the country's focus on digital transformation and economic diversification. It represents an evolving frontier within the realm of entrepreneurship, harnessing the potential of the Internet to transform traditional business practices and utilize digital technologies and the Internet to start, operate, and grow entrepreneurial ventures (Hasbolah et al., 2020). Based on its innovative nature, cyberpreneurship has emerged as a fundamental approach for enhancing entrepreneurial self-efficacy and strengthening the imperative for human capital development in shaping a resilient future (Newman et al., 2019).

Moreover, the rise of cyberpreneurship in Malaysia has sparked a phenomenon with the emergence of popular fashion brands like Bella Ammara, Zawara, Marshmallow Scarf, and others. These brands have gained significant recognition through various online platforms, such as websites, apps, marketplaces, and social media. Internet-based entrepreneurship is increasingly gaining popularity worldwide, including in Malaysia, as a means of creating and acquiring wealth. However, based on data collected by the Department of Statistics Malaysia in 2017 (Table 1), the adoption of information technologies, Internet usage, and web presence among Malaysian entrepreneurs remain relatively low, especially in the east coast states.

The Malaysia Digital SME Study in 2018 highlighted that 48% of SMEs in Malaysia faced challenges in digitalization primarily due to the lack of digital skills rather than financial or technological limitations (Malaysia Digital SME Study, SME Annual Report 2017/2018, 2018). Even though e-commerce adoption in Malaysia increased from 27.8% in 2019 to 37.9% in 2021, some entrepreneurs recorded preferring physical stores due to unfamiliarity with online business and e-commerce platforms and a lack of qualified employees for online system development and management (MSME, Insights 2021, Economic Performance and Outlook, 2021). Additionally, a joint survey by Adobe and Econsultancy, encompassing 13,000 professionals in marketing, creative, e-commerce, advertising, and IT worldwide, discovered that 43% of businesses in Asia, including Malaysia, faced digital skills and capabilities challenges stemming from workflow, legacy technology, and

Table 1 Cyberpreneurship to use technology adaptation practices among Malaysian business

States	ICT usage in business	Internet usage in business	Web presence usage in business
	73.5%	**61.7%**	**28.5%**
W.P. Kuala Lumpur	90.7	82.5	40.7
Selangor	87.9	78.5	43.1
Pulau Pinang	87.0	80.4	34.9
W.P. Putrajaya	83.9	76.9	36.3
Johor	74.7	60.6	24.1
Melaka	73.6	64.8	35.9
Perak	64.9	48.9	19.0
Pahang	64.8	47.2	18.9
W.P. Labuan	63.9	53.1	24.4
Sarawak	63.0	49.2	20.7
Kedah	59.3	45.6	17.8
Sabah	58.5	43.4	17.1
Perlis	52.7	42.0	16.9
Terengganu	51.0	36.8	13.7
Negeri Sembilan	50.4	38.8	14.9
Kelantan	46.5	32.1	11.5

Source: Department of Statistics Malaysia (2017)

systems issues (Tan, 2021). These facts raise a concern about the relatively slow adoption of emerging cyberpreneurship practices among businesses in Malaysia that may be impeding the success of the Fourth Industrial Revolution (4IR) era and Society 5.0 in Malaysia.

Digital marketing plays a vital role in cyberpreneurship by providing the strategies and techniques to reach and engage customers digitally effectively. Digital marketing offers a multichannel for consumers to control the buying process via product reviews and testing (Sidek et al., 2023). However, sales changes cause sales to become intricate without the proper marketing content to influence customers for sales prospects and repeat the sales cycle (Epstein, 2007). Even digital marketing requires high spending; however, top management often differentiates it from the traditional marketing channels with a good, established, and mature track record of return on investment (Leeflang et al., 2014).

Nowadays, the issues that come across in the digital marketing adaptation can be acknowledged based on prior research, which led to the conclusion that digital marketing benefits many consumers, but, at the same time, it is also problematic in its adaptation. Thus, it is causing difficulties for marketers to market their businesses in a competitive business setting. Hence, customer relationship management (CRM) can bridge the gap between information technology and marketing techniques to foster long-term partnerships and revenue (Glazer, 1997). CRM is a management method that consents businesses to attract, find, and retain worthwhile customers by dealing with their contacts (Hobby, 1999).

Digital media is currently recognized as the most powerful medium, and professional marketers may leverage this potential to improve consumer relationships (Neilpatel Homepage, 2022). It indicates that the level of CRM in digital marketing, which stimulates cyberpreneurship activities in Malaysia, needs attention and deep research. But unfortunately, there is little study on how customer relationship management accelerates digital marketing and directly promotes cyberpreneurship. As a result, this research will bridge the gap by analyzing the effect of CRM in stimulating cyberpreneurship activities in Malaysia by determining the nexus between pricing strategy, quality assurance, customer engagement, purchase intention, and CRM. The researchers believe that the pricing strategy, quality assurance, customer engagement, and purchase intention will directly affect CRM in stimulating cyberpreneurship in Malaysia. As a result, cyberpreneurship is evolving and becoming a current entrepreneurship trend, and digital marketing activities of an organization cover all marketing tactics.

2 Literature

2.1 Customer Relationship Management (CRM)

Customer relationship management (CRM) is one of the information management systems applications. Customers are an important link to the industry structure in a fast-growing industry. It is because customers are buyers of products or services a company produces. CRM has evolved into a collaborative, operational, and analytical three-dimensional structure and is a vital instrument in formulating cost-consciousness strategies for large-scale corporate assessment of customers (Buttle & Maklan, 2019).

Moreover, customers are committed to suppliers and make repeated purchases because they have built strong emotional connections. Emotions for customers may be difficult without compromising preferences and trust to deal with company complaints and problems effectively (Amofah & Ijaz, 2005). Abu-Alhaija et al. (2018) opined that customer loyalty is a long-term asset for a firm that seeks to lead its organization competitively by operating a solid CRM strategy. Moreover, the helpful experiences and principles inspire customers to invest impetuously as their affiliation with the organizational growth is consistent with high customer satisfaction. Additionally, digital technology enables companies to execute successful communication strategies and considers customer satisfaction and loyalty.

2.2 *Pricing Strategy and Customer Relationship Management*

The monetary value assigned to a product or service is determined by a mix of complicated mathematical study, knowledge, and willingness to take risks. The pricing strategy considers segments, ability to pay, market conditions, competitor actions, trading margins, and input costs. It is aimed at specific groups of customers and competitors (Leminen et al., 2006). For brands with a higher level of brand preference, retailers charge a premium and less consistent pricing, advertise more intensely, and coordinate pricing and promotions more closely. The most likely explanation for this pricing approach is that the brand has a "power of attraction" among consumers, meaning customers are very receptive when prices are reduced (Shang et al., 2007). Premium relative brand pricing, price inconsistencies, and price promotion changes are positively related to higher levels of relative brand advertising. These findings support the recent study by Tuti and Wicaksono (Tuti & Wicaksono, 2022); brand advertising is favorably connected to brand price and premium but adversely related to store profits. Based on this, the study believes pricing strategies have a significant positive relationship with customer relationship management. Therefore, the first hypothesis (***H1***) was formulated as follows:

H1: *There is a positively significant connection between pricing strategy and customer relationship management.*

2.3 *Quality Assurance and Customer Relationship Management*

Undeniably, Havlíček et al. (2013) categorize quality as a subcategory of marketing management, which is recognized in contemporary marketing trends as customer relationship management (CRM). Quality assurance topics have been extensively researched, not only limited to academic literature but also popular with practitioner-oriented publications. Quality assurance is an industrial method for managing and updating a quality system to continually guarantee and demonstrate that it meets specific requirements and standards (Vloeberghs & Bellens, 1996). The "Quality Management and Quality Assurance Standards" lays out the standards a manufacturer must meet to demonstrate its ability to plan, manufacture, and distribute goods or services with a coherent and consistent quality (Manghani, 2011). Quality assurance systems may turn into an entrepreneurial chance for growth and survival in an endlessly growing competitive market environment or as a restriction to business development, threatening them with contraction and marginalization (Pio et al., 2021). Accordingly, the current study predicts quality assurance may influence the potential of customer relationship management. Hence, the second hypothesis (***H2***) for this recent research was written as follows:

H2*: Quality assurance and customer relationship management have a significantly positive connection.*

2.4 Customer Engagement and Customer Relationship Management

It is vital to hypothesize and handle customer engagement to suit traditional marketing besides reaching current digital marketing trends. Customer engagement in commercial practices initially includes rational loyalty and emotional attachment (Appelbaum, 2022). Customer engagement is a psychological or emotional state of mind where the consumer is invested in a focused item, leading to regular interactions outside simple business motives (Rather et al., 2019). To Addo et al. (2021), customer engagement is beyond transaction and immediate purchase objectives, operated as an actual engagement hedonic or utilitarian with the focal item, product, or streamer. Visits, discussions, likes, and time spent interacting with the site are all examples of ways to engage. Users, users-hosts, and user platforms are all involved in this real-time synchronous interaction. The notion of marketing, which emphasizes satisfying customer wants, and the concept of market orientation, meeting customer demands by delivering greater value over rivals, both reflect a desire to bond with consumers. Moreover, customer engagement emphasizes offering long-term customer loyalty and understanding via greater value delivery, experiences, and products over their business rivals (Sashi, 2012). Thus, customer engagement is essential in developing commitment and confidence. Since then, this research aims to determine customer engagement, which can lead to customer relationship management. Henceforth, the study expressed the third hypothesis (***H3***):

H3*: Customer engagement and customer relationship management have a significantly positive connection.*

2.5 Purchase Intention and Customer Relationship Management

Purchase intention is a component of consumer cognitive behavior that describes how a person wants to purchase a specific product or service (Ling et al., 2010). It goes on to say that intention is more effective than behavioral measurements since customers may purchase due to restrictions rather than a genuine choice. Customers' purchase intentions were included in the current study, which clearly showed using buying purposes, including on live streaming pages. Instead of a simple perception as defined in previous studies, actual purchase intention counts the number of followers recorded by the stores, which were utilized to quantify followership and purchase intention, correspondingly (Algharabat, 2018; Cho & Sagynov, 2015;

Ghali-Zinoubi & Toukabri, 2019). Moreover, recent research has utilized the same metric. Since then, this research aims to determine purchase intention, which can lead to customer relationship management. Hence, this study formulated the fourth hypothesis (***H4***):

H4: *Purchase intention and customer relationship management have a significantly positive connection.*

3 Research Methods

This research aims to study the roles of CRM in stimulating cyberpreneurship in Malaysia by determining the nexus between pricing strategy, quality assurance, customer engagement, purchase intention, and CRM. The research was used as a quantitative study that employed a structured questionnaire as an instrument for the primary data collection. All variables in this study were used to build this research instrument. Meanwhile, the questionnaire was constructed in *Google Forms* and employed as a platform for questionnaire distribution, resulting in various advantages such as easier record management and time and cost savings. The variables measured the respondents' agreement level of a given statement with a five-point Likert scale: (1) strongly disagree, (2) disagree, (3) neither agree nor disagree, (4) agree, and (5) strongly agree. The summary of the items and measurements is depicted in Table 2.

Moreover, the non-probability, the purposive sampling technique, was chosen because this technique found the best appropriate sampling for this current study since the unit of analysis for this study is individual customers or online shoppers. At the same time, the sample was designed according to precise characteristics. The respondents in this study are individuals who engage in online shopping and must be above 18 years old due to this current study's main issue about customer relationship management. Further, the researchers assumed an individual under 18 years was not aware of the several terms used in the questionnaire. Also, referred to Krejcie and Morgan (1970), the suitable sample size for this study was 384 out of the total maximum population. Initially, 400 surveys were sent out to the targeted participants. After 8 months of conducting the survey, 375 responses from this study indicated 93.75% of the response rate. However, 35 samples have been omitted from the study due to incompleteness, the fact or state of not having any parts or being unfinished. Ultimately, data from 340 cross-sectional data were accepted and analyzed using descriptive, validity, reliability, normality, and multiple regression analysis through IBM SPSS Statistics software. The particulars of the main results and further analysis were elucidated in the result and discussion section.

Table 2 Measurement of variables

Variable/items	References
Customer relationship management (CRM)	Adikaram et al. (2016)
Seller promptly addresses complaints	
Seller consistently focuses on maintaining an appealing product	
Seller sends periodic email updates regarding new services	
Seller contacts regularly for personalized service	
Sellers highlight the significance of building lasting relationships	
Pricing strategy (PS)	Subrahmanyam et al. (2022)
Seller offers competitive prices to attract customers	
Seller provides affordable prices	
Seller sets prices based on the perceived customers value	
Seller offers customers the best deal possible	
Seller adjusts prices according to demand and competition	
Quality assurance (QA)	Chiguvi (2023) Laksamana (2018)
Seller always improves quality standards of product/service	
A seller consistently provides high-quality products/services	
Seller provides service platforms' security policy	
Seller continuously values customer feedback	
The seller offers warranties and guarantees of the product/service	
Customer engagement (CE)	Chu et al. (2020) Binsaeed et al. (2023)
I feel appreciated if the seller responds at a prompt rate	
Social media helps online customers get personal attention to their problems	
I feel appreciated if I can give opinions to other users	
I feel excited about product information and promotions on social media	
My recommendations on products and services are useful on social media	
Purchase intention	Laksamana (2018)
I have a great desire to start online purchases	
I intend to purchase online	
I am ready to purchase online	
I highly recommend online shopping to others	
I plan to keep purchasing online in the future	

Source: modified from many prior studies

4 Result and Discussions

4.1 The Descriptive Statistics

As stated in the preceding section, the sample was analyzed by using IBM SPSS Statistics software. It starts with a summary of the sociodemographic characteristics of the respondents, as shown in Table 3. Referring to this table, the study concluded

Table 3 Demographic characteristics of the respondents

Characteristics	Items	Frequency (n = 340)	Percentage (100.0%)
Gender	Female	217	63.8
	Male	123	36.2
Race	Malay	264	77.6
	Chinese	37	10.9
	Indian	31	9.1
	Others (*Bumiputera*)	8	2.4
Age (years old)	18–20	26	7.6
	21–30	272	80.0
	31–40	28	8.2
	41–50	8	2.4
	Above 51	6	1.8
Marital status	Married	134	39.4
	Single	198	58.2
	Single mother/father	8	2.4
Education level	Secondary	14	4.1
	Postsecondary education (*preuniversity*)	58	17.1
	Tertiary	265	77.9
	No formal education	3	0.9
Monthly average income (USD)[a]	Up to 100	68	20.0
	101–212	128	37.6
	212–423	49	14.4
	424–635	37	10.9
	636–847	32	9.4
	848–1058	9	2.6
	Above 1059	17	5.0

Source: Table established by the authors based on the 340 data output of the study
[a]Conversion rate is USD1 = RM4.61 as of June 19, 2023

that over 60% of the respondents were female compared to male. Additionally, over 77% of them were Malays, followed by Chinese (10.9%), Indian (9.1%), and others (2.4%). Furthermore, the data also showed that 95.9% were aged between 18 and 40 years, which concluded that most of the sample was youth, as defined by the National Youth Development Policy of Malaysia (aged between 15 and 40 years). Meanwhile, based on marital status, 58.2% were still single, while the remaining 39.4% were married, and 2.4% were single parents. Based on educational level, the data depicted almost 78% have completed their tertiary education; meanwhile, over 37% of them generate income worth USD212 (below RM1000).

Next, this study described the descriptive statistics summary as in Table 4. The descriptive statistics, expressing landscapes of a data set, was done by providing overall summaries regarding data samples. Overall, this study translated the measurement for all constructs as per the following scales: 1 to 2 were equivalent to a

Table 4 Descriptive statistics of the model

Variables	Mean	Standard deviations *(SD)*
Customer relationship management (CRM)	4.0853	0.68470
Pricing strategy (PS)	4.1012	0.66297
Quality assurance (QA)	4.2812	0.58978
Customer engagement (CE)	4.3153	0.61053
Purchase intention (PI)	4.3082	0.61356

Source: Table established by the authors based on the 340 data output of the study

Table 5 The reliability, validity, and normality analysis

Variables	Cronbach's alpha (α)	Skewness	Kurtosis
Customer relationship management (CRM)	0.859	−0.738	0.884
Pricing strategy (PS)	0.788	−0.449	−0.468
Quality assurance (QA)	0.838	−0.735	−0.018
Customer engagement (CE)	0.859	−0.825	0.130
Purchase intention (PI)	0.877	−0.727	−0.052

Source: Table established by the authors based on the 340 data output of the study

low agreement, 3 were narrated as moderate agreement, and 4 to 5 were measured as high agreement.

As portrayed in Table 4, all variables describe the mean values of all constructs above scale 4 (between 4.09 and 4.32), measured as the high agreement of the given statements. Thus, it specified that most of the 340 samples surpassed the high-level agreement on CRM of the main pillars: pricing strategy, quality assurances, customer engagement, and purchase intention.

4.2 *The Reliability, Validity, and Normality Analysis*

The reliability, validity, and normality analysis measures the validation and relationship of items in the variables. The summary result of the reliability, validity, and normality analysis of the model is illustrated in Table 5.

Firstly, the reliability analysis can be confirmed via Cronbach's alpha (α) value above 0.70. In Table 3, the range of Cronbach's alpha value for all variables was recorded between 0.788 and 0.877, which was higher than 0.70. Further, the variables' internal correctness concluded that the research notion had an appropriate standard. Similarly, (Cronbach, 1951; Bougie & Sekaran, 2020) also recommended that Cronbach's alpha value nearing the value 1 describes high internal consistency reliability. Therefore, the study achieved the reliability and validity analysis requirements.

After that, the normality analysis was determined by the value of skewness and kurtosis. Again, referring to Table 5, all variables in this research satisfied the value of −1.0 to +1.0 skewness and kurtosis, which concluded an adequate value to describe the normal distribution of the data. In conclusion, the data distribution of 340 samples in this study model was confirmed as normally distributed data. Hence, the multiple linear regression analysis was conducted to test all stated hypotheses. The discussions were detailed in the multiple linear regression analysis section.

4.3 The Multiple Linear Regression Analysis

This research aims to identify the level of pricing strategy, quality assurance, customer engagement, and purchase intention toward customer relationship management in digital marketing in Malaysia. The correlation (*R*) and coefficient of determination (*R*2) analysis were performed to determine the nexus among variables. According to Table 5, the model showed that the correlation between the CRM and the predicted CRM made by the model was 0.722. Additionally, 52.2% variation in the CRM can be explained by the pricing strategy, quality assurance, customer engagement, and purchase intention, while other factors can explain the remaining.

Next, the summary of hypothesis testing regarding the relationship between pricing strategy, quality assurance, customer engagement, and purchase intention on customer relationship management is presented in Table 6. Based on the result shown in Table 6, ***H1*** was supported by this study since the correlation value indicates a significantly positive association between pricing strategy and CRM. In other words, pricing strategy and CRM have a statistically significant linear relationship ($r = 0.484$, $p < 0.000$). Furthermore, it shows a positive direction of the relationship, where quality assurance was associated with CRM. Similarly, this research supports the ***H2***, where the relationship between quality assurance and customer relationship management was positively significant ($r = 0.371$, $p < 0.000$). Moreover, customer engagement and customer relationship management have a statistically positive link ($r = 0.220$, $p < 0.01$), as per stated in ***H3***. This indicates a significant positive interrelationship between customer engagement and customer relationship management. The last hypothesis, the ***H4***, also showed a positive

Table 6 The multiple linear regression analysis

R (R^2)	Hypothesis	Standardized beta	*t*-value	Significance level	Decision
0.722 (0.522)	***H1***	0.484	8.874	0.000	Significance
	H2	0.371	5.439	0.000	Significance
	H3	0.220	2.740	0.006	Significance
	H4	0.126	2.358	0.019	Significance

Note: Dependent variable (DV) is CRM. Source: Table established by the authors based on the 340 data output of the study

significant relationship between purchase intention and customer relationship management ($r = 0.126$, $p < 0.05$). Overall, all the formulated hypotheses were fully supported by the findings of many prior studies (refers Rather et al., 2019; Addo et al., 2021; Algharabat, 2018; Cho & Sagynov, 2015; Ghali-Zinoubi & Toukabri, 2019).

As per the discussion, companies need to implement communication strategies to satisfy customers with the services provided. Digital marketing strategy has many benefits; some firms are willing to pay large sums of money to advertise their companies online. The paradigm shift of tremendously increasing digital trends from 2020 to 2022 due to the COVID-19 pandemic has hit Malaysia and worldwide. As a result, digital marketing is evolving and becoming a current marketing trend.

Digital marketing plays a vital role in cyberpreneurship by providing the strategies and techniques to reach and engage customers digitally effectively. It helps cyberpreneurs effectively promote their products or services, build brand awareness, generate leads, drive website traffic, and convert prospects into paying customers. Additionally, digital marketing provides valuable insights and data analytics that cyberpreneurs can leverage to understand customer behavior, preferences, and trends. This information can assist in their business decisions, product development, and marketing strategies, leading to more targeted and effective digital marketing campaigns. Overall, cyberpreneurship relies on digital marketing as essential for establishing and growing a successful online business. It enables cyberpreneurs to leverage the power of digital channels, technologies, and data to connect with their target audience, drive business growth, and achieve their entrepreneurial goals.

The findings of this study highlighted the importance of incorporating CRM systems and digital marketing strategies into world policies such as Shared Prosperity Vision 2030 (SPV 2030) and Sustainable Development Goals (SDGs) and national or regional digital transformation policies such as MySTIE, MyDIGITAL, Industry4wrd, SME Digitalization Initiative, National E-Commerce Strategic Roadmap (NESR), etc. It can emphasize the need for businesses to adopt CRM technologies to optimize sales, marketing, and customer support functions, thereby contributing to the overall digital transformation goals of Malaysia and other countries.

5 Conclusion and Future Direction

The rapid rise of globalization and industrial modernization has impacted customer relationship management (CRM) in digital marketing. The practice of digital marketing in customer relationship management has influenced not only society but also countless organizations. Based on a recent survey, digital marketing knowledge is crucial because it has a significant impact, particularly in bringing Malaysia forward and making it a cashless society. Furthermore, CRM practices would be a competitive edge in stimulating digital marketing and as a conduit to collect important customer information, which can be used to enhance organizational growth.

Moreover, the traditional marketing approaches to increase customer retention are costly; hence, best CRM practices will offer firms a tactical strategy to treat this vital issue.

Despite the glowing result of findings, this current study has several limitations, and various past studies also highlighted future directions regarding this topic. First, the data in this recent research is solely based on cross-sectional data. Accordingly, longitudinal research would be satisfactory for new knowledge. Second, attention should be considered in the generalization of results, since this study employed non-probability sampling. Accordingly, various researchers recommended that empirical research using probability sampling would guarantee generalization and normality issues. Finally, future studies should consider other factors influencing CRM to refine the current model. The present concept is also limited because it ignores the importance of underpinning theory. Therefore, future studies should consider several theories, such as the motivational model (MM) (Davis et al., 1992), unified theory of acceptance and use of technology (UTAUT) (Venkatesh & Speier, 1999), technology acceptance model (TAM) (Davis, 1989), theory of planned behavior (TPB) (Ajzen, 1991), the model of personal computer utilization (MPCU) (Thompson et al., 1991), and innovation diffusion theory (IDT) (Rogers, 1995), to consolidate the conceptual framework.

Acknowledgments Recognition goes to Malaysia Ministry of Higher Education (KPT) as funder of this Fundamental Research Grant Scheme (FRGS) code **FRGS/1/2021/SS02/UMK/02/2**. Gratitude is also extended to all anonymous reviewers and editors who were directly or indirectly involved in the publication process of this article.

References

Abu-Alhaija, A. S., Raja Yusof, R. N., Hashim, H., & Jaharuddin, N. S. (2018). Determinants of customer loyalty: A review and future directions. *Australian Journal of Basic and Applied Sciences, 12*, 106–111.

Addo, P. C., Fang, J., Asare, A. O., & Kulbo, N. B. (2021). Customer engagement and purchase intention in live-streaming digital marketing platforms. *The Service Industries Journal, 41*(2), 1–20.

Adikaram, C. A. N. K., Khatibi, A., & Yajid, M. (2016). The relationship between customer relationship management and customer satisfaction: A study on private higher education institutions in Sri Lanka. *International Journal of Arts and Commerce, 5*(2), 69–95.

Ajzen, I. (1991). The theory of planned behavior. *Organizational Behavior and Human Decision Processes, 50*(2), 179–211.

Algharabat, R. S. (2018). The role of telepresence and user engagement in Co-creation value and purchase intention: Online retail context. *Journal of Internet Commerce, 17*(1), 1–25.

Amofah, P., & Ijaz, A. (2005). *Objectives, strategies and expected benefits of CRM*, 7-10.

Appelbaum, A. *The Constant Customer Gallup Business Journal*. Last accessed 2022/10/20, from https://news.gallup.com/businessjournal/745/constant-customer.aspx

Binsaeed, R. H., Yousaf, Z., Grigorescu, A., Chitescu, R. I., Nassani, A. A., & Samoila, A. (2023). Customer engagement and customer relationship management capabilities' effects on innovation performance and customer distrust's moderating role. *Sustainability, 15*, 9475.

Bougie, R., & Sekaran, U. (2020). *Research method for business: A skill building approach* (8th ed.). Wiley.
Buttle, F., & Maklan, S. (2019). *Customer relationship management: Concepts and technologies* (4th ed.). Routledge.
Chiguvi, D. (2023). Analysis of the effectiveness of e-customer service platforms on customer satisfaction at ABSA, Botswana. *International Journal of Research in Business and Social Science, 2147–4478*(12), 57–71.
Cho, Y. C., & Sagynov, E. (2015). Exploring factors that affect usefulness, ease of use, trust, and purchase intention in the online environment. *International Journal of Management & Information Systems (IJMIS), 19*(1), 21–36.
Chu, S. C., Chen, H. T., & Gan, C. (2020). Consumers' engagement with corporate social responsibility CSR communication in social media: Evidence from China and the United States. *Journal of Business Research, 110*, 260–271.
Cronbach, L. J. (1951). Coefficient alpha and the internal structure of tests. *Psychometrika, 16*, 297–334.
Davis, F. D. (1989). Perceived usefulness, perceived ease of use, and user acceptance of information technology. *MIS Quarterly, 13*(3), 318–339.
Davis, F., Bagozzi, R., & Warshaw, P. (1992). Extrinsic and intrinsic motivation to use computers in the workplace. *Journal of Applied Social Psychology, 22*(14), 1111–1132.
Epstein, M. J. (2007). Evaluating the effectiveness of internet marketing initiatives. In *Management accounting guideline*. The Society of Management Accountants of Canada, The American Institute of Certified Public Accountants and The Chartered Institute of Management Accountants.
Ghali-Zinoubi, Z., & Toukabri, M. (2019). The antecedents of the consumer purchase intention: Sensitivity to price and involvement in organic product: Moderating role of product regional identity. *Trends in Food Science and Technology, 90*, 175–179.
Glazer, R. (1997). Strategy and structure in information-intensive markets: The relationship between marketing and IT. *Journal of Market Focused Management, 2*(1), 65–81.
Hasbolah, H., Mamat, S. A., Abdullah, Z., & Sidek, S. (2020). A review on cyber entrepreneurship in Malaysia: Past, present, and future. *Journal of Critical Reviews, 14*, 2628–2636.
Havlíček, K., Břečková, P., & Zampeta, V. (2013). Quality management as a part of CRM. *European Research Studies, XVI* (SMEs), 15–27
Hobby, J. (1999). Looking after the one who matters. *Accountancy Age, 28*(October), 28–30.
Krejcie, R. V., & Morgan, D. W. (1970). Determining sample size for research activities. *Educational and Psychological Measurement, 30*, 607–610.
Laksamana, P. (2018). Impact of social media marketing on purchase intention and brand loyalty: Evidence from Indonesia's Banking Industry. *International Review of Management and Marketing, 8*(1), 13–18.
Leeflang, P. S., Verhoef, P. C., Dahlström, P., & Freundt, T. (2014). Challenges and solutions for marketing in a digital era. *European Management Journal, 32*(1), 1–12.
Leminen, S., Tinnilä, M., & Miikkulainen, K. (2006). Strategic pricing in business relationships - Do not miss the opportunity to create value for the customers. In: *The Annual ANZMAC conference*. Australia.
Ling, K. C., Chai, L. T., & Piew, T. H. (2010). The effects of shopping orientations, online trust and prior online purchase experience toward customers' online purchase intention. *International Business Research, 3*(3), 63–76.
Malaysia Digital SME Study, SME Annual Report 2017/2018. (2018). Accessed June 27, 2003, from https://www.smecorp.gov.my/images/SMEAR/SMEAR2017/ENG/Chapter2BoxArticle2.pdf
Manghani, K. (2011). Quality assurance: Importance of systems and standard operating procedures. *Perspective Clinical Research, 2*(1), 34–37.
MSME Insights 2021, Economic Performance and Outlook. (2021). Accessed June 27, 2023, from https://www.smecorp.gov.my/images/Publication/MSME_Insights/ekonomi/Economic%20Performance%20Outlook%202021%20-%20Box%20Article.pdf

Neilpatel Homepage, *Digital marketing made simple: A step-by-step guide*. Last accessed 2022/10/20, from https://neilpatel.com/what-is-digital-marketing/
Newman, A., Obschonka, M., Schwarz, S., Cohen, M., & Nielsen, I. (2019). Entrepreneurial self-efficacy: A systematic review of the literature on its theoretical foundations, measurement, antecedents, and outcomes, and an agenda for future research. *Journal of Vocational Behavior, 110*(October 2017), 403–419.
Pio, L., Cavaliere, L. P. L., Khan, R., Sundram, S., Jainani, K., Bagale, G., Chakravarthi, K., Rajan, R., & Rajest, S. (2021). The impact of customer relationship management on customer satisfaction and retention: The mediation of service quality. *Journal of Physiotherapy and Rehabilitation, 32*(3), 22107–22121.
Rather, R. A., Hollebeek, L. D., & Islam, J. U. (2019). Tourism-based customer engagement: The construct, antecedents, and consequences. *The Service Industries Journal, 39*(7–8), 519–540.
Rogers, E. M. (1995). *Diffusion of innovations* (4th ed.). The New York Free Press.
Sashi, C. M. (2012). Customer engagement, buyer-seller relationships, and social media. *Management Decision, 50*(2), 253–272.
Shang, W., Wu, H., & Ji, Z. (2007). Applying CRM in information product pricing. In L. Xu, A. Tjoa, & S. Chaudhry (Eds.), *IFIP international federation for information processing 2007* (Vol. 255, pp. 1407–1411). Springer.
Sidek, S., Hasbolah, H., Mohamed Khadri, N., & A., Mahshar, M., Abdullah, F., A., Mohd Rosli, M. (2023). Galvanizing digital marketing implementation in small businesses. In B. Alareeni & A. Hamdan (Eds.), *Impact of artificial intelligence, and the fourth industrial revolution on business success. ICBT 2021* (Lecture notes in networks and systems) (Vol. 485). Springer.
Subrahmanyam, S., Azad, F., & Arif, S. (2022). Premium pricing strategy and customer retention -An analysis. *International Journal of Early Childhood Special Education, 14*(03), 2064–2075.
Świeczak, W., & Łukowski, W. (2016). Lead generation strategy as a multichannel mechanism of growth of a modern enterprise. *Marketing Instytucji Naukowych i Badawczych, 3*(21), 105–140.
Tan, J. (2021). *Lack of digital skills and workflow issues hinder Asia marketers, finds Adobe study*. Accessed June 27, 2023, from, https://www.marketing-interactive.com/lack-of-digital-skills-and-workflow-issues-hinder-asia-marketers-finds-adobe-study
Thompson, R. L., Higgins, C. A., & Howell, J. M. (1991). Personal computing: Toward a conceptual model of utilization. *MIS Quarterly, 15*(1), 124–143.
Tuti, M., & Wicaksono, N. (2022). Customer relationship management and brand image's influence on loyalty through decision to stay. *Jurnal Bisnis dan Manajemen, 8*(2), 214–222.
Venkatesh, V., & Speier, C. (1999). Computer technology training in the workplace: A longitudinal investigation of the effect of the mood. *Organizational Behavior and Human Decision Processes, 79*(1), 1–28.
Vloeberghs, D., & Bellens, J. (1996). ISO 9000 in Belgium: Experience of Belgian quality managers and HRM. *European Management Journal, 14*(2), 207–211.

Syamsuriana Sidek is a Senior Lecturer at the Faculty of Entrepreneurship and Business, Universiti Malaysia Kelantan (UMK). She acts as project leader, first author, and corresponding author for this article. She mainly participates in the conception and planning of the writing article.

Hazrina Hasbolah is a Lecturer at the Faculty of Entrepreneurship and Business, Universiti Malaysia Kelantan (UMK). She acts as an expert in cyberpreneurship studies and networking with the respondents.

Nur Syafiqah A. Samad is a Senior Lecturer at the Faculty of Entrepreneurship and Business, Universiti Malaysia Kelantan. She acts as an expert in methodology and data analysis.

Zailani Abdullah is an Associate Professor at the Faculty of Entrepreneurship and Business, Universiti Malaysia Kelantan (UMK). He acts as a mentor of the project since he is expert in cyberpreneurship studies.

Nurul Husnina Najwa Zoraimi is a postgraduate student at the Faculty of Entrepreneurship and Business, Universiti Malaysia Kelantan (UMK). She acts as an expert in cyberpreneurship studies and networking with the potential respondents.

Nurul Azwa Mohamed Khadri is a Lecturer at the Faculty of Entrepreneurship and Business, Universiti Malaysia Kelantan. She acts as an expert in entrepreneurship theories.

Muhamad Ali Imran Kamarudin is a senior lecturer in Universiti Utara Malaysia (UUM). He is an expert in entrepreneurship specializing in family business, SMEs, entrepreneurial training and development as well as franchise business management. His area of interest is the development of family business SMEs, entrepreneurial training and development program and franchise business management studies.

Nor Hizami Hassin is a Lecturer at the Faculty of Earth Science, Universiti Malaysia Kelantan (UMK). He acts as an expert in economics studies and networking with potential respondents.

Husband's Supports Behind Successful Malay Women Entrepreneurs in Kelantan

Billy Kang and Mohd Rafi Yaacob

Abstract Women have always been praised for standing behind the successes of their men, but not the other way around. Due to various factors, including equal access to education, more and more women have an important and dominant role in socioeconomic activities and are more successful than their husbands. As a wife, a caring husband is essential to advancing a competitive woman, yet so little is being researched about his role. Against this background, this work investigates how husbands render the support and how do successful women entrepreneurs perceive the support as enhancing their success in their business. Altogether, five women entrepreneurs in Kelantan were selected for the study. The findings show that women entrepreneurs' support from their husbands helped sustain the wives' business aspirations. Data collected revealed that not much instrumental support was rendered other than occasionally helping to look after the shops. Emotional support was the main support provided by the husbands, and it came in the form of giving suggestions and options and sharing ideas about business. Egalitarianism, tradition, and culture of Kelantan influenced how the husbands rendered their support and how the wives perceived the support they received. Women entrepreneurs were grateful that their husbands allowed them to manage their businesses as they thought fit. In other words, their husbands' noninterference in their daily business operations is a blessing.

Keywords Husband · Support · Entrepreneurs · Success · Women · Culture

B. Kang · M. R. Yaacob (✉)
Universiti Malaysia Kelantan, Pengkalan Chepa, Kota Bharu, Kelantan, Malaysia
e-mail: rafi@umk.edu.my

N. Hayat, N. C. Nawi (eds.), *Contemporary Issues in Entrepreneurship and Innovative Technology*, Contributions to Environmental Sciences & Innovative Business Technology, https://doi.org/10.1007/978-3-031-50927-8_10

1 Introduction

It goes without saying that women have always been praised for standing behind the successes of their men. This social construct has been accepted since time immemorial, and there is an old, famous saying that "behind every successful man, there stands a woman," which has remained true until today – a wife who provides unconditional love and physical and emotional support for her husband. In synch with this, as far as social science research, including business and management research, is concerned, many studies have focused on the supportive role of the wives of males' successes as CEOs and entrepreneurs. Due to various reasons, among others, religious practices, equal access to education between men and women, women's empowerment, family backgrounds that place equal emphasis on both genders, and inclusive cultures that allow women to engage in entrepreneurship, more and more women have played the important and dominant role in socioeconomic activities.

In Malaysia, much research has been conducted on the typology of women entrepreneurs and their leadership styles among successful Malay women entrepreneurs (Fuad & Bohari, 2011; Hoe et al., 2012; Raman et al., 2008).

On the contrary, the information regarding the roles that the male spouse plays in the entrepreneurial wife's motivation and success is quite limited. It can happen in two situations. First, both a husband and a wife work, but the latter is more successful regarding salary and position. Second, a wife is the sole breadwinner, and her husband seemingly plays a less dominant role in earning the family's income. While husbands in both situations are still under research, the latter is more interesting to investigate as it challenges the status quo of the traditional role of husband and wife.

As a wife, a caring husband is essential to advancing a competitive woman, yet so little is being researched about his role. In the current gender equality world, husbands who are not working and rely on their wives as the main breadwinner are expected to share household chores during their wives' absence at home, as well as someone who listens to their wives and can share their opinions, as well as someone who values their opinions. They seemingly play a less important role than their wives. However, few studies have been conducted on the roles such husbands play.

Apart from looking at the roles of their husbands from the perspective of women entrepreneurs, the justification for this study can be looked at from another perspective; in that, it is not easy for supportive husbands to do what they are doing, such as sharing the house chores and "standing behind" their entrepreneur-wives, especially in the conservative Muslim Malay society of Kelantan. They are going against the grain of male chauvinism and patriarchal social norms. Therefore, it would be academically interesting to seek an explanation for the changing gender role phenomenon concerning the husband's behavior.

Jodyanne Jane Kirkwood (2009), a lecturer from Otago Polytechnic, used a qualitative lens to investigate spousal roles in motivations for entrepreneurship in New Zealand. Her study involved 68 entrepreneurs—45 men and 23 women. Regarding men, she posed an interesting question: "Are there invisible men behind the scenes who contribute to the business in the same ways that females do in male-owned businesses?". She found that spousal support has been observed to positively impact the entrepreneur's wife's success. Successful women would perceive their partners as more supportive, the biggest "fan" and "cheerleaders" (Cheung & Halpern, 2010). This contention is reinforced by Fuad and Bohari's (2011) study in Malaysia, and both argued that there was a positive correlation between entrepreneurial success and spouse support. Similarly, a previous study by Das (2001) in India found that a significant number of women attributed their successes to the support delivered by their spouses. In a further study, Slaughter (2014) suggested that "caregiver men are essential to the advancement of competitor women" and "that means, for every aspiring woman CEO, her chances of getting there will be considerably increased if her husband is the primary caregiver."

In Malaysia, Rajani and Sarada (Rajani & Sarada, 2008) found emotional support from spouses, family members, relatives, and other support systems. Other researchers in Malaysia, Nurwahidah and Abdul (2011), found that support positively correlated to entrepreneurial success. Family and partner support was recognized as playing an important role in women's achievement. The interconnectedness of social and family life characterizes women's leadership roles. Supports are important means for them to balance their social, personal, and professional lives. Work and family are not two distinct spheres but are more permeable in the case of women entrepreneurs than their male counterparts.

In this regard, the basis of this paper is to answer the questions: How do husbands render support? And how do successful women entrepreneurs perceive the support as enhancing their success in their business endeavors? Altogether, five women entrepreneurs were selected for the study.

2 Kelantan: The Land of Women Entrepreneurs

In Kelantan, women are very economically inclined, which can be clearly observed from the mainly female entrepreneurs in the main wet market of Siti Khadijah in Kota Bharu, the capital city of Kelantan. If one is walking at the said market, one will notice most of the traders are Malay women. It is not limited to the capital city, also observable in other districts' wet markets, and they also play key roles in many family-owned businesses (Mohd Rhouse, 2013; Rashid et al., 2015). Outside Kelantan, there are many successful women entrepreneurs. Two famous corporate ladies are Datuk Seri Hasmiza Othman of Qu Putih cosmetic products and the famed Naelofar Hijab by Noor Neelofa Bt Mohd Noor. There were still other successful Kelantanese women entrepreneurs, but they shied away from publicity.

Kelantan, being the cradle of Malay customs and culture, is in many ways still very conservative and has a traditional extended family structure, and family support is still very evident (Rahman, 2014; Hashim et al., 2012). Therefore, it is inconceivable that women can venture into business without their families' and husbands' implicit and explicit support.

They will inadvertently be seen as the entrepreneur breadwinners of the family (Nurwahidah & Abdul, 2011). Rashid, Ngah, Mohamed, and Mansor (Mohd Rhouse, 2013) noted that successful women entrepreneurs in Kelantan are self-reliant and have the ability to exploit opportunities, good people skills, risk-taking attitude. Ahmad (Hashim et al., 2012) suggests that Kelantan entrepreneurs are well known for their deep-rooted Malay identity and culture of business dealing. Due to the above qualities, the Malay women of Kelantan can reach their goals and be successful (Mohd Rhouse, 2013).

Due to the omnipresence of women in the business sphere in Kelantan, with little or no evidence of the proactive activities of their husbands in business in the eyes of visitors outside Kelantan, it comes as no surprise that their husbands, Kelantanese men, got a negative connotation as "lazy" or "indolence." This accusation is biased and oversimplified based on narrow, short observations lacking empirical support and rigor.

3 Common Criteria for the Selection of Women Entrepreneurs

Several criteria were used for the selection. First, the business had to be operated and managed by the wife. She had to be the wife, who was the key person operating and managing the business. Second, the business had been in operation for 5 years or more. The reason was that the business had gone through the "start-up" phase and moved onto the sustained growth stage. At this stage, the entrepreneur had to call on all her management skills and knowledge to maintain the smooth running of the business. At this stage, the euphoria of starting a new business had passed, and now working in the "real world" scenario. Under such a scenario, the demands of her time and work would reduce her time for the family, and she would need her husband's support to help maintain family harmony. Third, no fixed rule existed on the sources or designated pool of targeted samples. It was a purposive random sampling. However, it would be more cost-efficient if the target samples could be introduced by friends, bankers, or people who had direct contact with the targeted samples. It would solicit the targeted sample's willingness to be the research subject.

3.1 Background of the Respondents: Successful Female Entrepreneurs

Five women entrepreneurs were selected based on the criteria and parameters noted. Table 1 shows the demography of the respondents.

P1 was a graduate of Business Administration from the Universiti Utara Malaysia (UUM) about two decades ago. Upon graduation, she immediately jumped into the frozen food business, partnering with her elder sister, who had a Food Science degree, by taking over their father's business. One year later, she married and had four children. The business was primarily in the retailing and wholesaling of frozen food. She was also engaged in food catering for a school hostel. The business was doing well, attaining about RM12,000 daily sales, but the competition was tough. The sisters would spend 10–12 h daily in their business. It took too much of her time away from the family. So, after 15 years, she decided to sell off her frozen food business to spend more time with her family. But she could not stay idle for long; after less than a year's rest, she opened her fitness center business.

P2 was not highly educated, completed Form 6 (Upper Secondary), and her husband studied at Giat Mara, a vocational training school. Together, they had three children. Two were in college, and the youngest was completing Form 5 (Lower Secondary). She set up her frozen food business in 2007, upon the urging of her husband. Since then, she has expanded to three branches. Before setting up her present business, she was a shop assistant with a monthly salary of RM150, which she complained about by saying, "… eat food also not enough"! But she comforted herself that "never mind, I wanted to gain experience." Later, when her husband suggested the frozen food business, she agreed to try it. The rest is history. Her husband is an air-conditioning technician who works part-time because he does his work whenever his customers call in.

P3, who holds a Degree in Accounting from the UUM, set up the halal herbal beverage manufacturing business in 2012 with seed money of RM50,000 from her savings. As far as the said business was concerned, it was a small sum of money, but she had good relations with the suppliers from her former company, where she

Table 1 Demography of women entrepreneurs

	Participant	Age	Education level	Working experience prior to setting up own business (year)	Years as entrepreneur	Type of business
1	P1	38	Degree	0	16	Fitness center
2	P2	47	Secondary	5	12	Frozen food
3	P3	38	Degree	8	7	Halal herbal beverage
4	P4	46	Secondary	2	20	Bakery and mini market
5	P5	41	Degree	10	8	Halal cosmetic

worked for 8 years prior to resigning and starting her present company. Before setting up on her own, P3 was working for an SME company manufacturing halal beverage. She was good and dedicated and worked very hard, often putting in overtime. For that, she was rewarded and appointed to the position of Director of Administration. She presently owns her manufacturing plant of 20,000 sq. ft. at the PKINK Halal Food Industry hub at Pengkalan Chepa, with a loan from the Malaysian Industrial Development Finance Bhd (MIDF), and also set up a branch where her husband helped to look after it.

P4 did not have any tertiary education, completed Form 5, secondary schooling, and her husband attended trade school at Giat Mara. They were once burger sellers at the night market in Pasir Mas, Kelantan. They were overjoyed with the nightly revenue of about RM1000. In 1994, they saw an opportunity to take over a bakery in the town facing financial difficulty. However, when the negotiation did not go as planned, the couple decided to venture into burger bun making. At that time, they were actually buying the buns from an out-of-state supplier and saw the chance to fill the gap as a local burger bun supplier. With the meager sum of RM5000, they acquired a used dough mixer machine as seed capital. Two years later, they upgraded to a more efficient machine with the capacity to produce the equivalent of 100 bags of flour per day with the help of a small loan from a local bank. The higher production capacity meant they had to find a larger market. Indeed, they expanded their market coverage from their hometown of Pasir Mas to other cities like Kota Bharu, Tanah Merah, and Pasir Putih and even as far away as 60 km to Jerteh, the border town in Terengganu. Fast forward to 2018, they had built a new factory to house their bakery with three production lines producing a few kinds of pastry and bread. Burger bun remains their best seller. The couple also obtained the distributorship of Ramly Burger, the well-known Malaysian burger brand, for the whole state of Kelantan.

P5, 41 years old, hails from Bachok, Kelantan, and is a local university graduate majoring in Business Administration. P5 started her present business in 2011. Initially, she was mainly involved in the retailing of cosmetic products. In 2013, she rented a factory at Mara Entrepreneur Park, Pengkalan Chepa, Kelantan. With the new factory, she started her halal cosmetic products, such as foundation cake, face wash, toner, liquid foundation, and cosmetic products. Her products received certification from the Ministry of Health. Apart from producing her brand of products, she also acted as a consultant, giving advice on product formulation, packaging, and authority approval to other smaller-scale entrepreneurs without production facilities. "Like a one-stop center," she announced. The company also acted as the OEM producer for other, more well-known brands. P5 funded her operations with her initial capital of RM100,000, a combination of her savings and family contributions. In 2014, she doubled her production capacity by acquiring an adjacent factory. By then, she had a workforce of 28 staff members. P5's objective was to be a comprehensive and capable Bumiputra company targeted "to have halal products that can be marketed to the overseas markets," as she emphasized. P5 is married to a polygamous husband. He is also in business, so he would spend a lot of time looking after his business. In fact, when P5 completed her university education in 2003, she joined her husband's business – learning the ropes of entrepreneurship from her husband.

3.2 *The Supports Rendered by Husbands*

P1 expressed her satisfaction with the support rendered by her husband. When asked whether the husband would help out or not, she readily exclaimed, "Got … got. Always help out. No problem there." P2 appreciated that her husband did not "interfere with my business, but he comes and looks at my budget," and she was thankful that her husband had helped her to keep aside a small amount of cash "for a rainy day."

P1 and P3 were pleased that their husbands were "very understanding" of their entrepreneurial efforts. P4 showed her pleasure with her husband's contribution to her business endeavor and said, "He is very motivated … if there is a designation CEO … he is the one (she laughed)." She said her husband "has many ideas" even though she was the key person running the business. Similarly, P2 attributed the idea of setting up the frozen food business to her husband's suggestion. P5, who married a polygamous husband. Although she did not say much about her husband, she gave the impression that he did not stay with her "full-time," as she had described himself as a "part-time" husband. That being the case, she was observed to be very confident of herself and very independent in her business pursuits when she emphasized that she started "from zero" on her own. Could this be because her "man was not around" and she had no choice but to rely on herself?

When the question of whether they would discuss their problems with their husbands was encountered in their daily activities in running their businesses, apparently, none of the participants had a serious or formal discussion with their husbands! It reflected the confidence the wives had in themselves.

4 Discussion

The support women entrepreneurs received from their husbands essentially helped to sustain the wives' business aspirations. Data collected revealed that not much instrumental support was rendered other than occasionally helping to look after the shops. The data revealed that emotional support is the main support provided by the husband. This emotional support comes in the form of giving "suggestions," "options," and "sharing" of ideas, as informed by the husbands. Egalitarianism, tradition, and culture of Kelantan influenced how the husbands rendered their support and how the wives perceived the support they received.

This study is in line with other studies. It confirms the same study by Jodyanne Jane Kirkwood (2009) in New Zealand and Cheung and Halpern (2010) in the United States. The former found that spousal support positively impacted the success of the entrepreneur-wife, and the latter found that successful women perceived their partners to be more supportive of them. Similarly, in a previous study by Das (2001) in India, a significant number of women attributed their successes to the support delivered by their spouses. In the local scene, the study by Fuad and Bohari (2011) found a positive correlation between entrepreneurial success and spouse support. Other researchers, Nurwahidah and Abdul (2011) found the said support positively correlated to entrepreneurial success in Malaysia. Family and partner support plays an important role in women's achievement. The interconnectedness of social and family life characterizes women's leadership roles. Supports are important means for them to balance their social, personal, and professional lives. Work and family are not two distinct spheres but are more permeable in the case of women entrepreneurs than their male counterparts.

One observation stood out prominently in the way the wife managed her business. When was the question of whether they would discuss problems encountered in their business with the husband? Apparently, none of the participants had a serious or formal discussion with their husbands! It reflected the confidence the wives had in themselves. In order words, due to the business acumen they have acquired over the time since they embarked on their businesses, they can handle the nitty gritty of business themselves. Without their husband's helps and assistance, they managed to run their business successfully.

While it is not exaggerated to say that without their husbands, women entrepreneurs could run businesses themselves, they admitted that their husbands played a role in their business's sustainability. Having self-confidence alone is not enough to sustain their business inspiration. Their husbands' support is the key factor that impacted their perception and reinforced their perseverance. It might imply that support given by the husband is not necessarily related to the daily running business. Husbands who realize their wives have become capable of running their business need to look after it without no or less of their interference. And as things stand, the husband accepts and comes to terms with their arrangement, where they play what seemingly less important role in business as perceived by other people.

5 Conclusion

The support provided by successful women entrepreneurs in Kelantan from their husbands went beyond just emotional support. In addition to being understanding and encouraging, the husbands played a pivotal role in ensuring the success of their wives' businesses by offering practical assistance. This support was a reflection of the evolving dynamics of gender roles in the region. Practical support from husbands often involved taking care of household responsibilities, freeing up time and energy for the wives to focus on their businesses. Many husbands actively participated in childcare, household chores, and other familial duties, recognizing the significance of their wives' entrepreneurial endeavors and the positive impact it had on the family's overall well-being. Furthermore, husbands also acted as valuable advisors and strategists. They used their experience and knowledge to provide insightful suggestions, helping their wives navigate the challenges of running a business. With a collaborative approach, couples jointly discussed and brainstormed business ideas, fostering an environment of mutual respect and partnership. The egalitarian principles practiced within Kelantan's society played a crucial role in shaping these supportive relationships. Gender equality and mutual respect were highly valued, creating an atmosphere where both spouses worked together as a team to achieve their individual and collective goals. While tradition and culture have historically influenced gender roles in the region, the success of these women entrepreneurs demonstrated a shift toward a more progressive mindset. The community began recognizing and celebrating the contributions of women in business, challenging traditional stereotypes, and promoting inclusivity. In response to this support, the women felt empowered and appreciated the freedom given to them by their husbands to manage their businesses independently. The noninterference of their husbands in daily business operations allowed them to make decisions based on their expertise and intuition, fostering a sense of autonomy and self-reliance. Overall, the collaborative and supportive nature of these husband-wife partnerships has been a driving force behind the sustainable success of women-led businesses in Kelantan. As this trend continues to gain momentum, it has the potential to serve as a positive example for other communities, promoting gender equality and creating an environment where all individuals can thrive in their entrepreneurial pursuits.

References

Cheung, F. M., & Halpern, D. F. (2010). Women at the top: Powerful leaders define success as work+ family in a culture of gender. *American Psychologist, 65*(3), 182.

Das, M. (2001). Women entrepreneurs from India: Problems, motivations and success factors. *Journal of Small Business and Entrepreneurship, 15*(4).

Fuad, N., & Bohari, A. M. (2011). Malay women entrepreneurs in the small and medium-sized ICT-related business: A study on need for achievement. *International Journal of Business and Social Science, 2*(13).

Hashim, R. S., Ho-Abdullah, I., Raihanah, M. M., Yusof, N. M., Hamdan, S. I., & Jamsari, E. A. (2012). *Construction of a Muslim woman entrepreneur in Malay fiction.*

Hoe, C. H., Isa, F. M., Hin, C. W., Hashim, N., Yunus, J. M., & Abdullah, H. H. (2012). Development of women entrepreneurs: The case of Malaysia. *WORLD, 2*(6).

Kirkwood, J. (2009). Spousal roles on motivations for entrepreneurship: A qualitative study in New Zealand. *Journal of Family and Economic Issues, 30*(4), 372–385. https://doi.org/10.1007/s10834-009-9169-4

Mohd Rhouse, S. (2013). *Values and identities of women entrepreneurs: A study of Muslim women of Malay ethnicity in Malaysia* [PhD Thesis]. University of Waikato.

Nurwahidah, F., & Abdul, M. B. (2011). Malay women entrepreneurs in the small and medium-sized ICT-related business: A study on need for achievement. *International Journal of Business and Social Science, 2*(13).

Rahman, R. A. (2014). *The theory and practice of couples managing two full-time careers in Malaysia* [PhD Thesis]. Manchester Business School.

Rajani, N. and Sarada, D. "Women entrepreneurship and support system" 2008. www.krepublishers.com/02-Journal/

Raman, K., Anantharaman, R. N., & Jayasingam, S. (2008). Motivational factors affecting entrepreneurial decision: A comparison between Malaysian women entrepreneurs and women non-entrepreneurs. *Communications of the IBIMA, 2*(12), 85–89.

Rashid, K. M., Ngah, H. C., Mohamed, Z., & Mansor, N. (2015). Success factors among women entrepreneur in Malaysia. *International Academic Research Journal of Business and Technology, 1*(2), 28–36.

Slaughter, A.-M. (2014). Behind every great woman is a great man.pdf. *World Economic Forum Blog*. http://forumblog.org/2014/01/behind-every-great-woman-great-man/

Billy Kang has been in public practice for 30 years under the name of Billy Kang & Co PLT(AF 0783), specializing in assurance and compliance for small and medium-sized enterprises. He also provides his clients consultancy on financial and tax planning, business expansion, and acquisition advisory services. He operates in Cambodia under BG Associates Ltd. The firm provides cross-border assurance and compliance services for international clients.

Before setting up his public practice firm 30 years ago, Billy Kang trained in Hanafiah, Raslan, and Mohamed (later, Arthur Anderson). He later moved on to serve as Internal Auditor of a medium-sized Malaysian bank and later as General Manager of a property development company.

He is a Chartered Accountant of the Malaysia Institute of Accountants, an ACCA (UK) Fellow, and a KICPAA (Cambodia) member. Billy holds a Master's in Management degree from the Asian Institute of Management (AIM), Manila, and a Doctorate in Business Administration from the University Malaysia Kelantan (UMK). In the accountancy fraternity, Billy served as a Council Member of the Malaysian Institute of Accountants (MIA) (2008/11) and a Member of the ACCA Malaysia Advisory Committee (2008/9).

While in the Council of MIA, Billy chaired the Public Practice Committee, Audit Practice Sub-Committee, and Investigation Committee. Billy Kang is an Independent Director, chairing the Audit Committee for PPSEZ Ltd since 2014, a publicly listed company on the Cambodia Stock Exchange. Professionally, he is a regular speaker in the public forum on issues relating to Small and Medium-sized Practices (SMP), notably auditing standards and practices and quality control management.

Mohd Rafi Yaacob hails from Kota Bharu, Kelantan, and is a Professor of Entrepreneurship at Universiti Malaysia Kelantan (UMK). He grew up and spent his happy formative years near Wakaf Che Yeh, situated on the outskirts of the capital of the state, where, in those days, it was surrounded by rice fields, fruit orchards, small rubber farms, and bushes and had an early education in Sekolah Kebangsaan Kota and Sekolah Menengah Kota for secondary education, about two kilometers from his home. Later, pursued his studies at the Foundation of Science at Universiti Malaya in 1985. He earned a BSc (Geography) from Universiti Malaya in 1991 and an MBA from Middlesex University, UK 1996. To quench his thirst for knowledge, he pursued a Ph.D. at the University of Newcastle, Australia, in 2004 under the International Postgraduate Research Student (IPRS) from the said university and completed his study in 2007. He earned a Postgraduate Diploma in Entrepreneurship (PDGE) from the University of Cambridge, UK, in 2011. He began his academic career as a tutor at the Department of Geography, Universiti Malaya (1991–1994) and worked as a lecturer at the Faculty of Business and Management, Universiti Teknologi MARA (1997–2008). During his career at Universiti Malaysia Kelantan, he held various positions–head department, deputy dean, and dean of the Faculty of Entrepreneurship and Business. He was the university's Deputy Vice-Chancellor (Academic and International) from 2018 to 2021. He has been a member of the Malaysian Chartered Institute of Logistics and Transport (MCILT) since 2017 and the Malaysian Nature Society (MNS). His research interests are entrepreneurship, small business, strategic environmental management, and green business. He actively participates in research, publishes articles in refereed journals, presents papers at conferences, and has over a dozen books under his belt. Apart from reading, writing, and carpentering, he enjoys traveling, exploring the least traveled places to see, and spending time in the peace and quiet of unspoilt nature and culture.

Role of Family in Fostering Students' Sustainable Entrepreneurial Intentions

Kiran Kumar Thoti

Abstract Students have serious difficulties making career decisions in the present economic and social climate. Since the unemployment rate among college students may be effectively mitigated by entrepreneurial efforts, having an entrepreneurial mindset is crucial. Given this context, this article explores how families might best support kids in developing long-term, financially viable entrepreneurial goals. Twenty-five articles were selected from Scopus data after a careful search and evaluation. The study aimed to dissect how the model influenced the formation of hypotheses. The results demonstrate a strong correlation between kids' entrepreneurial aspirations and their families' business experiences and encouragement. It shows that these factors are crucial in encouraging long-term entrepreneurial motivation in pupils. The results also provided a novel understanding of how parental role modelling and financial literacy might inspire long-term entrepreneurial motivation in adolescents. Furthermore, the research suggests that entrepreneurship education be supported since it effectively stimulates student entrepreneurial intent.

Keywords Family role · Sustainable · Entrepreneurship · Students · Intentions · Entrepreneurial · Family support · Role model · Economics · Financial

1 Introduction

The present economic and social climate poses a significant obstacle for young people in acquiring and developing the needed skills in entrepreneurship. Young people must view entrepreneurship as a viable job option. People who start their businesses are sometimes called "innovators" or "pioneers" of new industries.

K. K. Thoti (✉)
Faculty of Entrepreneurship and Business, Universiti Malaysia Kelantan,
Pengkalan Chepa, Kota Bharu, Kelantan, Malaysia
e-mail: kiran.kt@umk.edu.my

N. Hayat, N. C. Nawi (eds.), *Contemporary Issues in Entrepreneurship and Innovative Technology*, Contributions to Environmental Sciences & Innovative Business Technology, https://doi.org/10.1007/978-3-031-50927-8_11

According to McCullough and Medina (2019), entrepreneurship may be defined as "the endeavour of creating, owning, and commercializing an idea, technology, product, or service, as well as taking on the risks and rewards associated with that enterprise." According to Haris Chechi (2023), entrepreneurship is significant in many facets of life, including a nation's social and economic elements. In their most basic form, entrepreneurs are the trailblazers who usher in the arrival of new technologies and processes, which, in turn, show in societal changes. These shifts are linked to improvements in lifestyle, more altruistic thinking, greater morale, and more economic options (Dave, 2020). When considering the significance of company ownership to the nation and the community, raising awareness about the significance of commerce to the community is of the utmost importance. Individuals with the ambition to engage in entrepreneurial activities by creating new businesses using current business possibilities are said to have entrepreneurial intentions. Individuals can have a good mindset and behave positively, considering numerous hazards associated with business when they have entrepreneurial ambition. Individuals with entrepreneurial purposes can conceive and create new commercial enterprises (Rita Pilar Romero-Galisteo et al., 2022). Students who have a high entrepreneurial intention have a positive attitude toward entrepreneurship, work hard to turn their entrepreneurial intention into entrepreneurial behavior, demonstrate entrepreneurial activities, are prepared to face the risks associated with entrepreneurship, and have the potential to launch a business venture in the future (Mark, 2023). In contrast, students who have a low entrepreneurial intention are more likely to have negative attitudes about the many dangers that one must face while engaging in entrepreneurial activity, to be confused when presented with possibilities to engage in entrepreneurial activity and to give up quickly when confronted with hurdles to engage in entrepreneurial activity. In a more general sense, low intention to be entrepreneurial leads to a lower employment rate.

Several elements, including social support, may influence the ambition to start a business. The feeling or the sense of receiving affection, care, value, and support from other people is what we refer to as social support. Family support is one of the many forms of social support, and research has shown that it is one factor influencing the desire to start a business. According to Algarme (2019), the term "family support" refers to the conduct and attitude of a family toward welcoming its family members. This acceptance might take the shape of emotional, informational, and instrumental assistance. As a result, I have concluded that one of the most significant factors that might inspire students to be interested in environmentally responsible business is their family.

1.1 Problem Statement

The problem of graduates being unable to find work is becoming more pressing in this age of globalization. Many nations indeed endure high rates of graduate unemployment each year. Still, if solutions to this issue are not discovered, our graduates' futures and our economy's viability may be in jeopardy. According to the figures, the unemployment rate in Malaysia will be 3.6% in December 2022, and there will be a total of 599,600 people without jobs. Salleh (2022) shows a considerable correlation between graduates' unemployment in Malaysia and low pay, job mismatch, quality of higher education systems in Malaysia, and graduates' employability skills. In addition, *The Straits Times* news asserts that the effect of graduates' unemployment leads to inefficiencies in using available resources. The expansion of the national economy may be hampered if recent graduates find employment in fields unrelated to their areas of expertise. According to the Department of Statistics Malaysia, most graduates in Malaysia come from households that fall into the B40 socioeconomic class, and these graduates report having a difficult time finding work after completing their education. If these problems continue, future graduates may be at risk of poverty.

These difficulties must be addressed so that this problem may be contained and handled explicitly for the nation to reach the required development level. Due to its inability to satisfy the requirements of economic development necessary to achieve the status of a developed nation, Malaysia is classified as a developing country. To accomplish this objective, finding a solution to the unemployment issue among graduate students is essential. Considering this, one of the most efficient ways to solve this issue is to encourage young people, particularly recent graduates, to pursue careers in business. It is starting a business not only has the potential to help solve the issue of unemployment among graduates but also has the potential to enhance national revenue indirectly. I feel that the role of the family is significant to give birth to the student's goal to be a sustainable entrepreneurial model. As a result, this research was carried out to investigate the most significant parts of the family that play a role in the likelihood of our nation's population developing an entrepreneurial spirit in the years to come.

2 Literature Review

The second chapter's objective is to provide a literature review to contribute to the existing body of information about the function of the family in the development of students' long-term intentions to start their sustainable businesses. It incorporates the statement or hypothesis developed based on the results of the prior article. This chapter will touch on various definitions of the significant characteristics covered in the present research: family entrepreneurial history, family support, role modelling, financial literacy, and birth order in the family. Various meanings of these primary components will be touched on throughout this chapter. A research framework is also addressed, so be sure not to forget about it.

According to Amirul et al.'s research from 2023, the home environment, particularly the parents, has a significant impact in influencing the entrepreneurial goals of kids, which in turn may be one of the professional paths students choose in the future. Family members have the potential to be considered role models and have a big effect on the desire to start a business. Not only do aspiring business owners have successful businesspeople to look up to as models, but they also have these successful entrepreneurs' financial and emotional support. Because of this, the home environment influences the pupils' intent to engage in entrepreneurial activity. Both the help of the university and the support of the student's family favorably influence the student's attitude toward entrepreneurship; however, the perception of family support has statistically more substantial consequences than the perception of university support. According to Purwana et al.'s research from 2023, students' good attitudes about entrepreneurship positively influence their entrepreneurial intent, and students' positive experiences with entrepreneurial pursuits have an affirmative effect on their entrepreneurial intention. In addition, children and teenagers look up to their parents as role models, hoping they will emulate their achievements. In addition, the families counsel their members on the value of working for themselves and give opportunities for them to do so. According to the findings of Ntharika Chilenga et al. (2022), one can conclude that people whose families have a history of entrepreneurial endeavors tend to have more significant self-employment goals than their contemporaries do.

Next, emotional support from family members might be linked to a signal of approbation, which is essential for individuals raised in collectivistic cultures and could help young people start their businesses. The provision of guidance is another kind of support that may be offered. It is common knowledge that parents significantly impact the choices of future careers made by their children who have just graduated from high school. According to research by Martin Mabunda Baluku et al. from 2020, this results in a considerable effect of family support on the transition from entrepreneurial goals to entrepreneurial activity. In addition, fear of failure has a strong negative influence on entrepreneurial aspirations, whereas entrepreneurial ability is seen to have a positive impact on entrepreneurial ambitions. The informal support that people get helps reduce the unfavorable association between the fear of failing and the initiative to start a business. According to Lejla Turulja et al.'s research from 2020, the presence of support from family and friends attenuates the unfavorable association between the fear of failing and the ambition to engage in entrepreneurial activity.

Additionally, the family's economic condition significantly improves entrepreneurial intention since it provides financial stability. Most students lack the financial money and resources necessary to launch an entrepreneurial venture, which poses barriers to entrance. According to Liu et al. (2022), one of the essential strategies for managing financial challenges faced by entrepreneurs, particularly younger entrepreneurs, is to seek aid from relatives. Additionally, parental support for autonomy, encouragement of independence, and exposure to an entrepreneurial role model are the most crucial elements that inspire the ambition to start a business. According to Moussa and 2021's research, the assistance that young people get from their families in financial and social capital has little impact on their desire to start their businesses. In addition, developing this form of connection from a young age may lead to a higher significance of the parental role model and a more substantial interest in the company being passed down from generation to generation (Gianni Roman et al., 2022). The following research question guides the study:

(a) This study aims to know the role of family support in fostering students' sustainable entrepreneurship intention.
(b) This study aims to evaluate the effect of family role modelling in fostering students' intention in sustainable entrepreneurship.
(c) This study evaluates financial literacy in developing students' sustainable entrepreneurship intention.
(d) This study aims to measure how well the birth order in the family affects students' sustainable entrepreneurship intention.

3 Research Method

A thorough search of academic databases like Scopus, including freshly published works, will be conducted. The search approach will include the keywords family (family), sustainable enterprise (SE), student intention (SI), and role (role). The first search is anticipated to yield several publications, according to an expected count of 225 items from Scopus. The retrieved articles will be selected based on predetermined inclusion and exclusion criteria. The inclusion requirements may be met by peer-reviewed English-language studies examining how families influence students' aspirations for sustainable entrepreneurship. Research that doesn't particularly addressed how families help children develop their SEI won't be considered. Data Extraction: The selected publications will be mined for details like authors, publication year, research objectives and methodology, key results, and implications. We will create a consistent data extraction form for consistency and accuracy. Based on preset criteria, including the suitability of the study subject, the appropriateness of the method, and the appropriateness of the data analysis, the quality and relevancy of the selected articles will be assessed. This assessment will help to ensure that the meta-analysis includes only articles of the highest quality. Analysis of Data: The acquired data will be analyzed using a theme analysis approach. This study aims to identify and synthesize key findings and issues regarding the birth order in families, the role of the family in fostering students' sustainable entrepreneurial intention, and its applications to student intention in entrepreneurship. It also examines the impact of family entrepreneurial background on students' intentions. The research questions will be addressed in integrating and reporting the findings. The review will include key applications, advantages, challenges, and recommendations for further study and development, as well as the function of families in promoting students' desire to engage in sustainable entrepreneurship. The results will be presented in an understandable and structured manner (Fig. 1).

The most often cited database is Scopus, which enables scholars to easily identify reliable research and access information that might guide them in choosing the best publication for their research paper. The market for research publications has significantly expanded, and almost daily new journals are being published. Identifying and removing low-performing and predatory journals from the list is necessary to select the best papers. Only well-known, effective, high-impact, and peer-reviewed publications are included in the Scopus database of journals (Borgel, 2020) (Fig. 2).

For the first search, keywords such as “artificial intelligence in agriculture” were utilized. However, there were a lot of false-positive findings. After several iterations, the literature was searched using a PRISMA-style four-stage methodology (Recommended Measures for Systematic Guidelines for Reviews and Meta-Analyses) using the query string TITLE (“Artificial Intelligence”). About 25 publications with the keywords Artificial Intelligence in Agriculture were utilized for the study, all from Scopus. The Scopus website returned 225 articles appropriate for “Artificial Intelligence in Agriculture” and had the incidence. They said that the VOS Viewer included 11 papers with the same incident. A Scopus database literature search revealed the record, which had a count of 127. The count with the final evaluations reached 25 after categorizing the study’s years and files into categories acceptable for data processing. The co-occurrence of the keywords and terms with general publishing trends and citation metrics.

4 Results and Discussions

The collected data was used to analyze and provide bibliometric qualities, including publishing and citation by year, publication, the top 10 most active source titles, and the top 10 most influential authors.

Network model based on the Bibliographic – Authors.

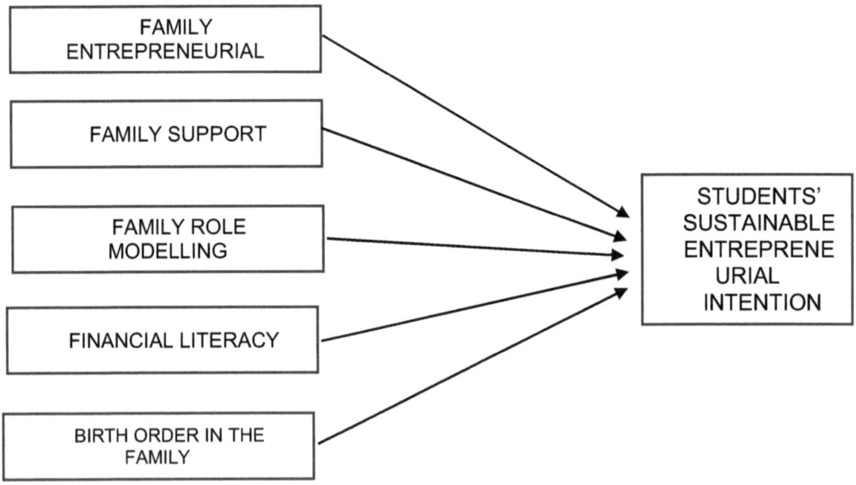

Fig. 1 Research model

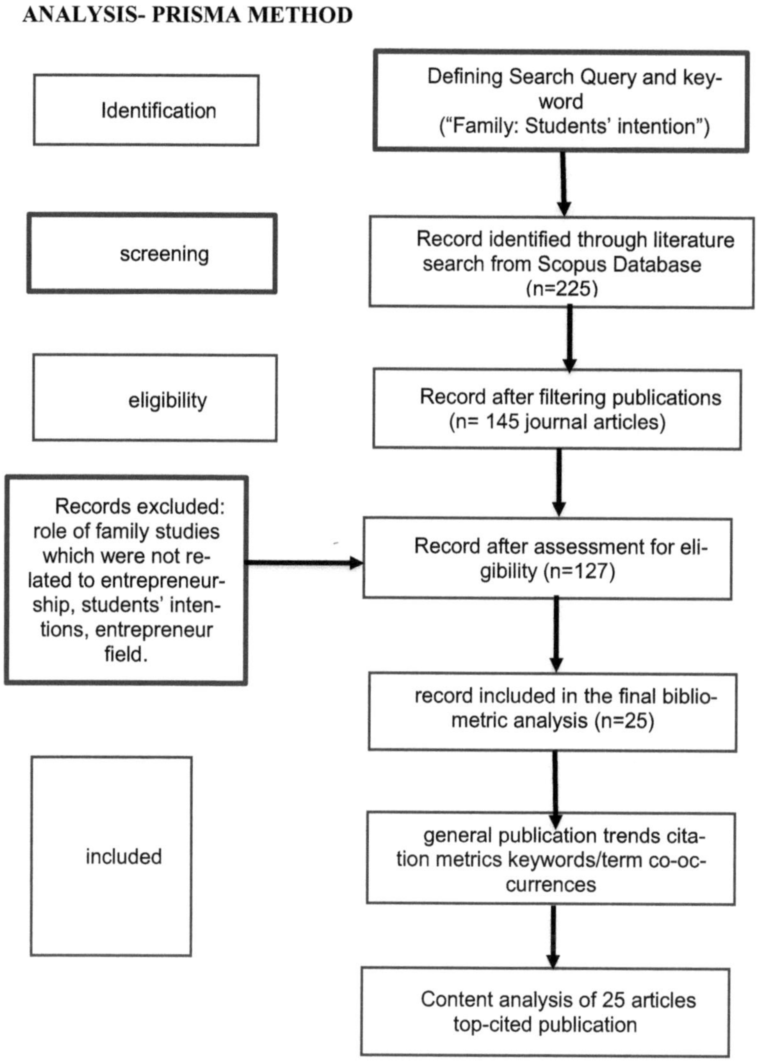

Fig. 2 PRISMA framework

Table 1 lists the top 20 authors and links to their relative strengths.
Network model with keywords or occurrences (Table 2).
Network model created based on text data – titles and abstracts.

ID	Term occurrence	Relevance score
1	Association	21.7526
2	Beginning	20.6663
3	Cognition	31.7721

Table 1 The top 20 authors and links to their relative strengths

ID	Author	Documents	Total link strength
8	"bibić, l.i."	1	13
16	"dragin, a.s."	1	13
17	"džigurski, a.i."	1	13
29	"huang, y."	1	4
32	"ivanović, o.m."	1	13
35	"knežević, m.n."	1	13
36	"košić, k."	1	13
43	"ma, z."	1	4
45	"mei, h."	1	4
46	"mijatov, m.b."	1	13
54	"ning, w."	1	4
70	"stankov, u."	1	13
71	"stoiljković, a."	1	13
72	"stojanović, v."	1	13
76	"tomić, s."	1	13
78	"vujičić, m.d."	1	13
79	"vuković, a.j."	1	13
80	"wang, j."	1	4
84	"wibowo, a."	2	0
92	"ðerčan, b."	1	13

ID	Term occurrence	Relevance score
4	Economic education	50.7587
5	Educational support	30.9448
6	Enterprising family	31.7721
7	Entrepreneurial outcome	21.7526
8	Entrepreneurship training	20
9	Existence	31.0338
10	First business	20.7587
11	Graduate	40.9448
12	Physical activity	30.659
13	Physical education	20.7587
14	Relational support	40.9525
15	Risk	20.9394
16	Romania	20.7587
17	Sex	31.0338
18	Socioeconomic level	21.0401
19	Sport	30.7566
20	Uncertainty	0.9448

Figure 3 shows a network model based on the Bibliographic by authors. Data were downloaded in ris format, and authors were credited for their contributions to

Table 2 Top 20 keywords

ID	Keyword	Occurrences	Total link strength
3	Parent	1	14
5	Article	1	14
11	Controlled study	1	14
13	Cross-sectional study	1	14
17	Economics	1	14
18	Education	1	14
24	Entrepreneur	4	10
30	Entrepreneurial intention	13	9
31	Entrepreneurial intentions	5	20
40	Entrepreneurship	6	17
41	Entrepreneurship education	5	8
54	Female	1	14
58	Gender	1	14
59	Gender inequality	1	14
67	Human	1	14
68	Human experiment	1	14
73	Kruskal-Wallis test	1	14
105	Self-efficacy	3	16
114	Student	5	12
115	Students	3	8

the study's design, findings, and interpretations. For each of the 92 authors, determine the aggregate strength of their coauthorship relationships with all other authors by selecting an arbitrary number of authors. Eighty-seven authors will be chosen based on their total link strength, which will be the ones reviewed.

Figure 4 shows the network model with keywords or occurrences. The data was obtained in ris format, and co-occurrence and keyword analysis were performed using complete counting as the counting method. Of the 129 keywords, 129 are relevant enough to fulfil the minimum selection criteria of 5 documents per author. Each of the 129 keywords will have its total strength of co-occurrence links with other keywords calculated, and the keywords with the highest total link strength will be selected based on the number of keywords you provide. The most vital and robust link keywords are parent, article, controlled study, cross-sectional study, economics, and education.

Next, in Fig. 5, the information was retrieved in ris format, and we have extracted the title and abstract fields from the keywords and a complete count. The selected cutoff value for the lowest frequency with which a particular word occurs is 2. Out of a total of 129, only 122 are acceptable. A relevance score will be determined for each of the 122 terms. The highest-scoring terms are chosen for further consideration. The default option uses a relevance score of 60% to pick the best phrases. There are 122 potential terms to choose from. The strongest and most potent link

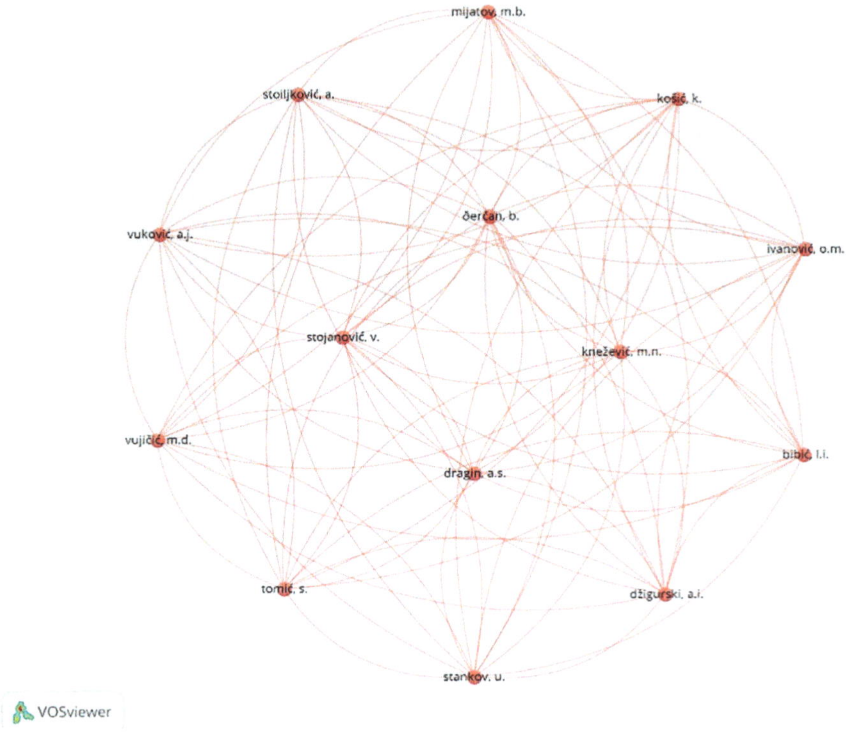

Fig. 3 Network model based on the Bibliographic – Authors

keywords are association, beginning, cognition, economic education, educational support, and enterprising family.

5 Conclusions and Recommendation

This research aimed to determine whether students' families are essential in developing their long-term intentions to start their businesses. We demonstrated that family dynamics play a varied role in determining whether young people are motivated to start a project of their own. Actually, the ambition to start a business is influenced by four aspects of a person's home environment. These aspects include parental support for autonomy, an entrepreneurial role model, family support for entrepreneurship, including financial resources and social capital, and family support for entrepreneurship. In addition, the family business history variable goes one step further. It reveals how parental exposure, family member exposure, and job engagement each has a separate effect on the elements that influence one's attitude toward entrepreneurship. Even while the findings indicate that entrepreneurial self-efficacy, attitude, and subjective norms all mediate the connection between one's family

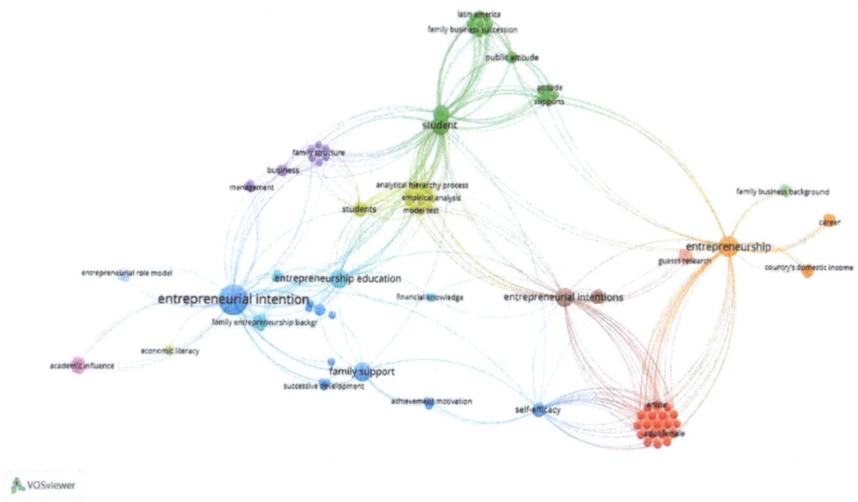

Fig. 4 Network model with keywords or occurrences

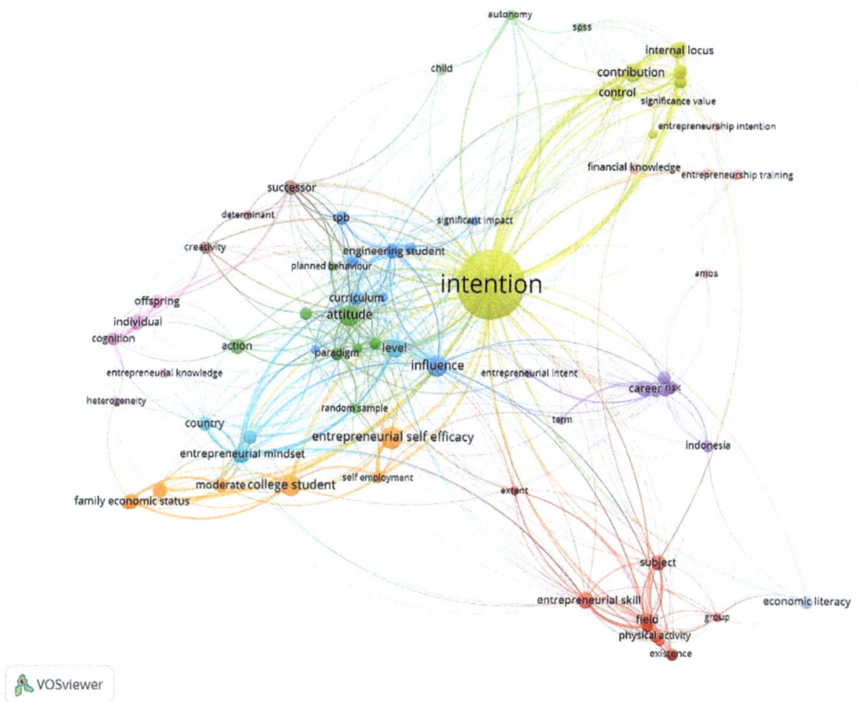

Fig. 5 Network model created based on text data – titles and abstracts

company past and their desire to put that background into practice, the strength of this attitudinal mediation depends on the specifics of one's family business background. The study's conclusion includes a discussion of the implications of the study's findings, both theoretical and practical, as well as the inherent limits of the study and the prospects for more research.

To successfully instill the purpose of succession in their offspring, parents who run family companies need to exercise more caution when dealing with the emotive component of the equation. Because they have a more significant effect on motivation and, as a result, affect the desire of the successor to remain in the firm, the emotional ties to the family business need to be created as early as possible in a person's life. Suppose this sense of connection to the organization is successfully awakened. In that case, it will probably be counterbalanced by the detrimental influence of the parental role model. Communicating with their children about their experiences as business owners and the positives and negatives of being an entrepreneur, as well as the satisfactions and dissatisfactions associated with being an entrepreneur, is essential for parents who run their businesses and want their children to take an interest in continuing the family enterprise.

Young people need to be aware of the factors identified as crucial to becoming a successor for them to continue with the companies their parents started. If these practices are not maintained, there is a potential that the business may be put out of business rather than contributing to the organization's consolidation and expansion. Young brains armed with fresh information gained during their time at university and in their professional training can help their parents' enterprises expand and prosper. There is no question that the succession of businesses by young professionals may be essential in enhancing and modernizing the organization and bringing it to a new level of output and profitability. In this view, it is necessary to create situations that will inspire or sensitize young professionals to use the knowledge they have gained at university in the family business, ultimately leading to an increase in the company's growth, competitiveness, and overall development. Many people who enroll in colleges and universities are the first members of their families to get a college degree. As a result, they were able to elevate the level of professionalism inside the family company.

Education in business ownership and management is an example of one of these examples. In addition to lessons on entrepreneurship, educational institutions must include in their curriculum sessions on managing family businesses. As far as we know, these classes are only offered at the graduate level, whereas the undergraduate level offers very few, if any, of them. Undergraduate students who take a course on family businesses might benefit from raising their awareness regarding the significance of these companies to the economic growth of their communities, regions, and nations. It would be especially true for those students who come from entrepreneurial families. Their university education could assist their parents' enterprises in becoming more competitive, creating more employment opportunities, and expanding from a microenterprise to a medium or big corporation.

Along these same lines, it is a good idea to provide examples of family businesses that have either been successful or unsuccessful. It may help the audience become more aware of the topic and prompt them to consider whether it would be appropriate for them to run the company their parents started one day. Students might be inspired to have a different perspective on their family companies and be more motivated to succeed if family company owners, both men and women, were invited to speak about their business experiences, including their successes and failures. Both male and female students might find themselves more interested in taking over the family company because of the incidents described here.

References

Algarme, C. (2019, September 2). *EO Melbourne - Entrepreneurs' Organization*. Retrieved July 17, 2023, from EO Melbourne - Entrepreneurs' Organization website: https://www.eomelbourne.org/blog/family-support-in-the-entrepreneurial-journey

Amirul, M., Eryanto, H., Syafira, N., Arif, A., Michael, A., & Wibowo, A. (2023). Evaluating the Structural Effect of Family Support and Entrepreneurship Training on Entrepreneurship Intention Among Indonesian University Students. *International Journal of Educational Methodology, 9*(1), 227–236. https://doi.org/10.12973/ijem.9.1.227

Baluku, M. M., Kikooma, J. F., Otto, K., König, C. J., & Nida. (2020). Positive psychological attributes and entrepreneurial intention and action: The moderating role of perceived family support. *Frontiers in Phychology, 11*, 546745. https://doi.org/10.3389/fpsyg.2020.546745

Borgel, N. (2020, Nov 07). *Why Scopus is Important*. Retrieved from econtentpro: https://www.econtentpro.com/blog/why-scopus-is-important/185

Chechi, H.. (2023, April 26). *Why is entrepreneurship important?* Retrieved July 17, 2023, from Glion Website: https://www.glion.edu/magazine/entrepreneurship-importance/#:~:text=Entrepreneurship%20is%20frequently%20credited%20as,problems%20while%20creating%20new%20products

Chilenga, N., Dhliwayo, S., & Chebo, A. K. (2022). The entrepreneurial mindset and self-employment intention of high school learners: The moderating role of family business ownership. *Frontiers in Education*, 7. https://doi.org/10.3389/feduc.2022.946389

Dave, M. (2020, March 30). *Top 5 takeaways on the importance of entrepreneurship - DCID*. Retrieved July 17, 2023, from DCID website: https://dcid.sanford.duke.edu/importance-of-entrepreneurship/

Liu, Y., Minteer, S. D., Li, X., & Zeng, J. (2022). Entrepreneurship education on entrepreneurial intention: The moderating role of the personality and family economic status. *Frontiers in Psychology, 13*, 1–2. https://doi.org/10.3389/fpsyg.2022.978480

Mark. (2023, March 26). *The 8 key characteristics of successful intrapreneurs - Intrapreneur nation*. Retrieved July 17, 2023, from Intrapreneur Nation website: https://intrapreneurnation.com/skills/8-characteristics-of-successful-intrapreneurs/#:~:text=Intrapreneurs%20are%20always%20open%20to,the%20boundaries%20of%20what's%20possible.

McCullough, D., & Medina, B. (2019, September 20). *What is entrepreneurship?* | Doug McCullough, Brooke Medina. Retrieved July 17, 2023, from Fee.org website: https://fee.org/articles/what-is-entrepreneurship/#:~:text=Simply%20put%2C%20entrepreneurship%20is%20the,rewards%20associated%20with%20that%20enterprise

Purwana, D., Widyastuti, U., & Respati, D. K. (2023). What drives the students' entrepreneurial intentions? Investigating the role of financial socialization and entrepreneurial education. *WSEAS Transactions on Business and Economics, 20*, 1144–1152. https://doi.org/10.37394/23207.2023.20.102

Romaní, G., Soria-Barreto, K., Honores-Marin, G., Roberto, R., & Rueda, J. (2022). Not Like My Parents! *The Intention to Become a Successor of Latin American Students with Entrepreneur Parents, 14*(3), 1193–1193. https://doi.org/10.3390/su14031193

Romero-Galisteo, R. P., Gonzalez-Sanchez, M., Gálvez-Ruiz, P., Palomo-Carrión, R., Casuso-Holgado, M. J., & Pinero-Pinto, E. (2022). Entrepreneurial intention, expectations of success and self-efficacy in undergraduate students of health sciences. *BMC Medical Education, 22*(1). https://doi.org/10.1186/s12909-022-03731-x

Salleh, B. N. (2022, June). *Factors influencing unemployment among graduates in Malaysia.* Retrieved from KMC uni. Razak: https://kmc.unirazak.edu.my/unemployment-among-graduates-in-malaysia-during-the-covid-19-era/#:~:text=Results%20have%20shown%20that%20factors,us%20via%20the%20button%20below

Turulja, L., Veselinović, L., Agić, E., & Pasic-Mesihovic, A. (2020). Entrepreneurial intention of students in Bosnia and Herzegovina: what type of support matters? *Journal: Economic Research-Ekonomska Istraživanja, 33*(1), 2713–2732. https://doi.org/10.1080/1331677x.2020.173021

Kiran Kumar Thoti Ph.D. in Management and expertise in Entrepreneurship, Human Resources, Information Systems, and Marketing Management. Have 15 years of Academic, Research, and Administrator Experience in Various Business Schools, Public Universities, Private Universities, and Institutions in India and Malaysia. I have a successful track in Research with articles published in 107 articles in Scopus, Web of Science, Google Scholar Index, UGC, and peer-reviewed journals in national and international journals and a decade of experience in administration: Head of the department, Chairman, Professor-in-Charge, Student Mobility Center Head, Academic Collaborations, Program Structure Design, Motivating Staff and Student toward progressive Development in the Academic Records. Academics have 15 years of experience teaching and creating courses in General Management, Human Resource Management, Information Systems, and Marketing Management for degree students, executive education students, online students, master students, and doctoral students. Guiding 06 PhD scholars/students, 10 master's students, and 10 bachelor students in India and Malaysia. One PhD scholar submitted the PhD thesis report, 102 Master students submitted, and 100 Bachelor students submitted and completed their academic projects. Organized two international conferences and two funding projects in progress. I believe knowledge is sharing and updating by learning new things.

Exploring the Kelantanese Youth Understanding Toward Microentrepreneurship

Kiran Kumar Thoti

Abstract The research objective was focused on the possible business opportunities for the young generation entrepreneurs in Kelantan, one of the states in Malaysia. The youth in Malaysia focuses their career on small-scale businesses, like food trucks, delivery services, finance services, textiles, and beauty clinics in Kelantan. The break-even point in the small-scale business sustainability was very difficult to manage. The youth won't consider the real world and are in the imagination of high profits in the initial stages of the business, in that few of the young generation entrepreneurs are successes, and most of them are drop out of their small-scale businesses. The research focused on the possible opportunities for the youth in Kelantan to adopt the microentrepreneur through survey methods, and the data was analyzed with the percentage method.

Keywords Micro business · Entrepreneurship · Risk management

1 Introduction

Small and medium-sized enterprises (SMEs) have been one of the community's professional possibilities, and it has also elevated in importance across a range of government initiatives. The government does acknowledge and commit to educational initiatives. After 1995, the Ministry of Entrepreneur Development (KPU) was established to foster interest in business entrepreneurship among young people and the youth, including the entire community. This initiative and training program strive to achieve this goal (Kerajaan negeri Bantu belia Bina kerjaya, 2022).

In addition, these young managers lack expertise and know-how to manage a firm's money; this issue frequently arises. When this group lacks patience, it results

K. K. Thoti (✉)
Universiti Malaysia Kelantan, Pengkalan Chepa, Kota Bharu, Kelantan, Malaysia
e-mail: kiran.kt@umk.edu.my

N. Hayat, N. C. Nawi (eds.), *Contemporary Issues in Entrepreneurship and Innovative Technology*, Contributions to Environmental Sciences & Innovative Business Technology, https://doi.org/10.1007/978-3-031-50927-8_12

in a lack of experience. They are eager to launch their own company. A handful of them lack the mindset to seek out experience. These people do not work hard to achieve their goals to help their firm expand successfully.

The United Nations General Assembly sets the age range for the youth generation as being between 19 and 29 years old. However, one of the government's priorities in the National Transformation (TN50) to prepare Malaysia as a developed and high-income country is the age level of the youth, which is calculated in Malaysia from the age of 15 to 40 (Belia Bahasa Melayu, ensiklopedia bebas, 2022). The Malaysian National Entrepreneurship Policy empowers the youth to develop the entrepreneurial mindset with the government support. Malaysian government offers support in all major sector to facilitate the youth to uptake the entrepreneurial venture. However, risk management reasons a difficult area as it requires skills and training to the youth to take the informed decisions to take the marketing and financial risks in best manner. Entrepreneurial failure is highly associated with the lack of financial risk management skills among the youth.

2 Literature Review

The handling of business risk among young microentrepreneurs in Kelantan is the main emphasis of this study. The Malaysian government is taking measures to encourage young people to engage in entrepreneurial activities by offering a 20 million grant initiative because youth entrepreneurship is not uncommon. It could be because of the perception that young people have original and inventive solutions for resolving social or environmental issues. The fact that these young entrepreneurs have the stamina and vigor to carry out their entrepreneurial and business endeavors makes them a group recognized as agents of economic change in the modern world (Yaacob et al., 2021). The number of young entrepreneurs is rising daily, thanks to their growing self-awareness of the state of the economy and their understanding of the advantages of being an entrepreneur who has the power to alter someone's course in life if they are skilled at spotting opportunities and openings. It is consistent with the National Digital Policy's goal of developing digital entrepreneurs who align with the skills and interests of the current young generation and the Ministry of International Trade and Industry of Malaysia's (2012) encouragement to produce many young entrepreneurs. This circumstance affects many young individuals who want to start great businesses and are interested in doing so.

Young people's involvement as entrepreneurs, the majority of whom are 77% in micro, small, and medium-sized firms, can lessen their reliance on the labor market as it is currently structured (Khan et al., 2016). Malaysia's Chief Statistician, Dato' Sri Dr Mohd Uzir Mahidin, commented on the report, saying: "Traders continue to be optimistic about business prospects for the second quarter of 2022, albeit at a slower pace, with confidence indicators recorded at +3.5% compared to +7.6% in the previous quarter. This circumstance demonstrates that young people are interested in entrepreneurship." Most young people would remain jobless after completing their degrees in their disciplines due to the intense rivalry in other professional domains. They are motivated by this scenario to start their businesses to survive.

Additionally, the government has allotted several funds to support young people in expanding this entrepreneurial sector. The state of Kelantan offers specialized haircutting classes for men and women that businesspeople run in the industry as part of the government's ongoing commitment to implementing various programs. The Industrial Job Placement Program 2022 will occur in Kuala Lumpur for 6 months. This event, which has the full backing of the state government of Kelantan, is held to assist young people in evaluating the skills and enterprises they wish to start (Fraser et al., 2022).

The National Entrepreneurship Policy is a program the government has improved to help young people interested in starting their businesses develop their attitude (DKN 2030)(Menon, 2022). As part of its assistance initiatives, the government would also target youth, Bumiputera, and other target groups in key, high-impact sectors such as smart agriculture, halal, aerospace, biotechnology, medical devices, and the digital economy. Using cutting-edge and digital technologies, business owners will be inspired to keep innovating in terms of management and operations (Razuan et al., 2019).

While doing this, the government will construct a supportive and all-encompassing ecosystem for entrepreneurship that considers the variety of traits that entrepreneurs, cooperatives, agriculture-based groups, social enterprises, and small and medium micro companies can have (SMEs). The government will implement policies to strengthen governance, rules, institutional roles, and financing to increase entrepreneur performance (Khan et al., 2016). Risk management issues that frequently arise involve marketing and financial damage. These young businesspeople are innovative and have fresh concepts for their products. But they are unsure of their strategy or formula to increase sales. Due to their improper marketing strategies and decreased visibility because of this predicament, the things they make are less well known. This issue is quite concerning because while many young people enter the field of entrepreneurship, some young people also experience financial failure because of it. The purpose of this work is to describe the degree and application of business risk management among young entrepreneurs. This paper aims to close the knowledge gap in Malaysian youth management literature and practice. This issue is crucial for the nation since the youth are the future and the engine of national development. Problem Statement: When young people enter these small and medium-sized enterprises (SMEs), inadequate financial risk management, such as failing to return every loan obtained, is a common issue (Khan et al., 2016). Entrepreneurs typically receive funding or loans from various sources, whether from the private sector or direct government assistance. This loan is provided to help business owners launch their ventures and can ease the load of start-up cash. The typical young entrepreneur irresponsibly uses borrowed money. For instance, they spend the loan money on items with a lower return. The loan money melts because of this circumstance. When the time for repayment comes, their income or profits from their firm cannot support the loan. Because of this circumstance, their names are placed on a blacklist for future loans. Any organization that might aid in this entrepreneurial sector will have doubts about the entrepreneur's skills when such a situation arises. Due to uncertainty and names on blocklists, this group will likely have trouble getting future loans and financial aid from any agency. It is because many young entrepreneurs lack direction and sound financial understanding. Every industry you want to enter has its requirements. Most of these young businesspeople disregard the necessity of expertise. One of the main factors in a young person's business failing is their inherent desire for achievement in a short amount of time. The youth's rate of thought is another factor contributing to this circumstance. Many youths engaged in this entrepreneurship have recently earned their Malaysian Certificate of Education (SPM). This group typically shows less interest in learning new things, so they frequently fail in business (Fraser et al., 2022).

The main objective of this research is to investigate the level of practice business of small-scale businesses among the youth. Still, it will be focused on the eastern state of Malaysia. The specific objective of the studies is to

- Understand the possible business opportunities for the youth or young generation to start small-scale businesses in Kelantan.

This study focuses on the extent and practice of business opportunities among young entrepreneurs in Kelantan. This study will be done with a soft copy questionnaire as a survey and reference for the researcher. This study will involve young entrepreneurs, both men and women, who are in Kelantan. This study also involves urban and rural areas to get a comprehensive overview of entrepreneurs' business risk management practices.

This research allows researchers to understand better what factors influence business risk management practices among youth entrepreneurs. The elements that affect business risk management can have a bad effect if entrepreneurs do not pay attention. In addition, this study also discusses the extent to which young entrepreneurs are familiar with business risk management practices. It is because of the objective research to examine the business possibilities implemented by micro-sized entrepreneurs. Many business risk management practices can be used, but not all entrepreneurs implement those methods. Among them is providing financial records of their business. This study examines the level and practice of young entrepreneurs toward starting new start-ups.

3 Method

Analysis Plan The primary and secondary data used in the research studies will be described in this section. This section addresses the study's goals and evaluates the research's speculation. The research's data have been examined, and this chapter presents the statistical analysis findings. The research study's entire set of findings is presented in tables. Descriptive statistics and Pearson correlation serve as the primary analytical tools in this research study. The researchers compiled, edited, classified, and encoded the data from the questionnaires using computer programs like the Statistical Package for the Social Sciences (SPSS), which they then used to interpret the primary data. With the aid of this computer program, researchers can quickly compute the data and motivate quantitative research in a less taxing manner. The variables are analyzed using descriptive analysis, and the respondent's biographical details are also described. Reliability statistics are used to determine the significance of the relationship between the study's variables.

Descriptive Statistics Descriptive statistics have been used to describe the key components of the data in an investigation. It provides simple summaries of the sample and the measurements. In descriptive statistics, the goal is to explain the significance of the data or what they will reveal. Researchers can use descriptive statistics to predict and visualize future data changes. It has been used to display the current quantitative descriptions in an approachable manner. It can help the researcher effectively simplify a large amount of data. The average value of each variable is calculated using descriptive statistics analysis.

4 Data Analysis and Discussion

4.1 Respondent Profile and Business Background

Table 1 provides a summary of the study participants' profiles. The study's findings indicate that the respondents' genders are nearly equal, with 50.9% women and 49.1% men. The age range of most respondents in this youth category was found to be between 25 and 34 years old (76.8%). If we examine the level of education, most respondents (42.9%) have a degree, followed by those who have at least an SPM/SPVM education (21.4%) and those who have a diploma (16.1%). The level of education typically has little bearing on how well a micro business is run because it typically relies on family inheritance, local resources, and the local market. In fact, it has become one of the primary requirements for receiving microfinancing assistance (Mohamad Syahmi, 2020). The study's findings also revealed that while most respondents only make between RM1001 and RM2500 per month (48.2%), the income of respondents can reach RM10,000 per month (4.5%). It is most likely connected to how the coronavirus, which has started to be found in Malaysia during this study, is expanding (Table 2).

Business Background as an Independent Variable An overview of the respondent's business data can be found in Table 3. According to the study's findings, most respondents (32.1%) work in the catering/restaurant industry, provide food services, and prepare food. According to the analysis, most respondents who work in the food industry make an average monthly income of between RM2501 and RM5000. A small warong or kiosk like Medan Selera can provide information about the location and size of the food business. Micro-food entrepreneurs can still sell on a small

Table 1 Respondent profile and business background

Demography	Item	Respondent	Percentage (%)
Gender	Male	55	49.1
	Female	57	50.9
Age	Below 25 years	13	11.6
	25–29 years	45	40.2
	30–34 years	41	36.6
	35–40 years	13	11.6
Level of education	Malaysian Certificate of Education (SPM)	27	24.1
	STPM, STAM, skills certificates	14	12.5
	Diploma	18	16.1
	Bachelor's degree	48	42.9
	Master's degree/doctor of philosophy (PhD)	5	4.5
Monthly income	RM1000 and below	12	10.7
	RM1001–RM2500	54	48.2
	RM2501–RM5000	36	32.1
	RM5001–RM7500	5	4.5
	RM7501–RM10,000	4	3.6
	RM10,0001 and above	1	0.9
Total		112	100

scale using kiosks or small stalls provided with a certain rental rate, despite selling in small, medium, or mega supermarkets. Small booths or kiosks are also available for other types of businesses, including the sale of communication devices, handicrafts, clothing and fabrics, cosmetics, and pharmaceuticals, among others. The government has adopted this situation to boost the income of micro-status business owners in Malaysia's major cities. The results show that 20.5% of the respondents work in service-related businesses, such as working at kiosks for communication tools, printing services, salon services, spa massages, and cosmetics, among other businesses. The analysis also revealed that respondents working in this service-related industry make an average monthly income of between RM1001 and RM5000.

Moreover, Table 3 indicates that the respondent's experience as an entrepreneur is in the range of 3 to 5 years (41.07%), followed by 6 to 8 years (25.69%). However, most respondents (61.6%) claimed their company had only been operating for 1–5 years. Despite having a business age of 1–5 years, 61.6% of the respondents, the analysis shows that 40.2% cannot hire employees. At the same time, 15.2% of respondents and 36.6% of respondents employ two employees to support their business operations.

Table 2 Respondent business background information

Independent variables	Category	Number of respondents	Percentage (%)
Type of business	Retails	20	17.9
	Food and beverage	36	32.1
	Handicraft	2	1.8
	Services	23	20.5
	Tourism	6	5.4
	Manufacturing	2	1.8
	Agriculture	8	7.1
	Clothes	15	13.4
Number of employees	No employees	45	40.2
	One person	41	36.6
	Two persons	17	15.2
	Three persons or above	9	8.03
Experience as an entrepreneur	1–2 years	27	24.11
	3–5 years	46	41.07
	6–8 years	29	25.89
	9–10 years	9	8.03
	Over 11 years	1	0.9
Business age	Below 1 years	26	22.3
	1–5 years	69	61.6
	6–10 years	17	15.2
	10 years and above	0	0
Total		112	100

Table 3 Dependent variable information

Dependent variable	Answers	Number of respondents	Percentage (%)
Microentrepreneurship (ME)	Yes, they understand the ME	97	86.7
	No, they don't understand ME	15	13.3
Total		112	100
Top factors and practices in business that entrepreneurs have picked	Do a financing record	48	42.8
	Recruit special staff for risk management. Take an insurance policy	12	10.7
	Make a financial control	17	15.2
	Use professional financial services. Updating market information (up and down the price of goods)	6 4 25	5.4 3.6 22.3
	Others	–	–
Total		112	100

4.2 Understanding the Microentrepreneurship

According to Table 3, 86.7% of respondents know microentrepreneurship. 13.3% of respondents said they knew little about microentrepreneurship. The situation is improving because, according to a study by Zulnaidi (2016), 42.8% of microentrepreneurs in Kelantan are unaware of or have limited knowledge of microentrepreneurship, which is a warning sign that needs to be addressed immediately. It relates to the business standing of small and medium-sized enterprises (Zulnaidi, 2018). Further investigation reveals that the respondents appear to use various business risk management techniques, some of which they may not be aware are part of standard business management procedures and which they have been using for a considerable amount of time. The respondents were given a list of business risk management procedures to choose from as one of their commercial affairs' most crucial risk management procedures. The table reveals that 42.8% of the respondents identified preparing and maintaining financial records as their company's primary risk management strategy. However, the information in the table does not imply that the respondent's company only uses one microentrepreneurship strategy.

5 Conclusion

This study's main aim is to discover the practices carried out in microentrepreneurship implemented in Kelantan. Among the microentrepreneurship practices in business are financial records, updating market information, appointing special staff for risk management, taking out insurance policies, financial control, using professional financial services, and others. This study was carried out with support from observation and review of several journals, articles related to this topic, and findings from microentrepreneurs in Kelantan. As a result, this study makes us more knowledgeable for all practical purposes, as stated in the main objective of our research. This study helps to improve the process of identifying practices in business risk management. It can open the eyes of the public, especially entrepreneurs in Kelantan, to know about business risk management which is increasingly playing an important role in affecting business performance in recent times. The survey findings confirm that the practice in business risk management implemented by microentrepreneurs in Kelantan is at a moderate rate and still needs to be improved and should not be taken lightly. Among the practices often implemented is preparing financial record books and updating market information. Other practices that are considered important are buying insurance policies, controlling one's finances, and having dedicated employees, especially in financial risk management for the business. Because this study only focuses on entrepreneurs at the microlevel, the use of professional financial services for business financial replacement is very rare due to the small size of the business.

Most respondents agreed that microentrepreneurship can provide great benefits to their business, especially in its financial performance and operations. Each business will be measured and assessed for its the level of risk experienced by referring to several indicators of the probability of the impact of various risks and bringing multiple effects to the business. In achieving business goals, whether small or big dreams, the chances that occur need to be faced and overcome concerning the level of impact they create. Each factor was studied and questioned, and all the relationships between practices by microentrepreneurs in Kelantan were analyzed using the correct method. However, more studies need to be conducted to get a more comprehensive picture of business risk management among entrepreneurs, especially after the outbreak of the COVID-19 pandemic that hit the world. It is because the business management of small entrepreneurs has begun to venture into new norms with the use of mobile and online technology that enables the existence of new forms of issues such as cybercrime and security. Youth microentrepreneurs need to be given serious attention for their survival and impact in influencing business performance now and in the future. This study will not cover other issues that are not necessarily related to the entrepreneur's perspective but only cover the practices the researcher has listed. This study will not involve business conditions before the COVID-19 pandemic, 2019, until 2020. Respondents must be micro-sized entrepreneurs who reside and conduct business in Kelantan.

References

Belia Bahasa Melayu, ensiklopedia bebas. (2022). *Wikipedia*, ensiklopedia bebas. Retrieved January 8, 2023, from https://ms.wikipedia.org/wiki/Belia#:~:text=Istilah%20%22belia%22%20secara%20umumnya%20merujuk,berumur%2016%20hingga%2032%20tahun

Fraser, J. R. S., Quail, R., & Simkins, B. J. (2022). *Questions asked about enterprise risk management by risk practitioners* (pp. 251–260). *Business Horizons*.

Kerajaan negeri Bantu belia Bina kerjaya. (2022, February 10). *HarakahDaily*. https://harakahdaily.net/index.php/2022/02/10/kerajaan-negeri-bantu-belia-bina-kerjaya/

Khan, S. J. M., Nur Syamilah, M., & Noor, A. R. (2016). Performance of youth entrepreneurs in Malaysia micro small and medium enterprises. *International Journal Of Economics and Financial Issues, 6*(S7), 86–88.

Menon, S. (2022). More Malaysian youths venturing into business amid challenging market, *The Star.* https://www.thestar.com.my/news/education/2022/08/14/fortune-favoursthe-bold

Razuan, M. R., Zaimah, R., Sarmila, M. S., & Awang, A. H. (2019). Kualiti keusahawanan dalam kalangan usahawan kecil Dan sederhana. *Malaysian Journal of Society and Space, 15*(3). https://doi.org/10.17576/geo-2019-1503-02

Yaacob, M. H., Thing, N. S., & Alias, N. (2021). Information asymmetry and industry 4.0 among small and medium enterprises (SME) in Malaysian halal industry. *Jurnal Pengurusan, 63*, 31–49. https://doi.org/10.17576/pengurusan-2021-63-03

Kiran Kumar Thoti Ph.D. in Management and expertise in Entrepreneurship, Human Resources, Information Systems, and Marketing Management. Have 15 years of Academic, Research, and Administrator Experience in Various Business Schools, Public Universities, Private Universities, and Institutions in India and Malaysia. I have a successful track in Research with articles published in 107 articles in Scopus, Web of Science, Google Scholar Index, UGC, and peer-reviewed journals in national and international journals and a decade of experience in administration: Head of the department, Chairman, Professor-in-Charge, Student Mobility Center Head, Academic Collaborations, Program Structure Design, Motivating Staff and Student toward progressive Development in the Academic Records. Academics have 15 years of experience teaching and creating courses in General Management, Human Resource Management, Information Systems, and Marketing Management for degree students, executive education students, online students, master students, and doctoral students. Guiding 06 PhD scholars/students, 10 master's students, and 10 bachelor students in India and Malaysia. One PhD scholar submitted the PhD thesis report, 102 Master students submitted, and 100 Bachelor students submitted and completed their academic projects. Organized two international conferences and two funding projects in progress. I believe knowledge is sharing and updating by learning new things.

Empowering Non-educated Women Entrepreneurs in Kelantan for Digital Finance Inclusion

N. Fadzlyn and Nik Maheran Nik Muhammad

Abstract The Malaysian state of Kelantan boasts a vibrant entrepreneurial landscape, with women at the forefront, driving economic growth through their businesses. However, many women entrepreneurs face a common obstacle: a lack of formal education, leaving them inadequately equipped to navigate the evolving economy and technological advancements. Financial illiteracy, particularly concerning informal business practices, restricts their potential to achieve economies of scale and hinders their ability to make informed financial decisions, thus impeding business growth and expansion. Despite these challenges, the spirit of entrepreneurship and determination to provide for their families has driven many women to become primary breadwinners. This paper highlights the critical issue of financial literacy among women business owners in Kelantan and its impact on their capacity to access finance, secure loans, and make sound financial decisions. As many of these women entrepreneurs operate informal businesses, their growth potential is limited, hindering their ability to contribute substantially to the economy. To foster a healthy entrepreneurial ecosystem, targeted financial literacy programs must be developed to equip women entrepreneurs with the necessary knowledge and skills to manage their finances effectively, enabling long-term growth and financial inclusion. In conclusion, this chapter emphasizes women entrepreneurs' challenges due to financial exclusion and proposes solutions to help them adapt to digital finance. Empowering women through financial literacy, mentorship, and supportive policies will contribute to a thriving entrepreneurial ecosystem in Kelantan and foster sustainable economic growth.

Keywords Finance · Digital · Literacy · Women · Entrepreneurial · Inclusion

---

N. Fadzlyn (✉) · N. M. N. Muhammad
Universiti Malaysia Kelantan, Kota Bharu, Kelantan, Malaysia

N. Hayat, N. C. Nawi (eds.), *Contemporary Issues in Entrepreneurship and Innovative Technology*, Contributions to Environmental Sciences & Innovative Business Technology, https://doi.org/10.1007/978-3-031-50927-8_13

1 Introduction

The Malaysian state of Kelantan has a thriving entrepreneurial landscape, with many women driving economic growth through their businesses (Sultana et al., 2015). Women entrepreneurs play an essential role in providing for their families and communities. However, a substantial number of these women business owners share a common difficulty: they are inadequately educated about the economy and technological advances since they lack formal education. Most of these businesswomen are illiterate in financial matters related to informal business practices, which limits their ability to achieve economies of scale. Despite these challenges, the entrepreneurial spirit and will to live have driven many of them to become the primary breadwinners in their families. Nevertheless, a lack of financial awareness is a significant challenge for women business owners in Kelantan. Even though many of these dedicated women are incredibly entrepreneurial, they often lack the financial literacy to run their businesses successfully. The gap in financial literacy limits their capacity to acquire finance, secure loans, and make informed financial decisions, stifling their enterprises' growth and expansion (Wang et al., 2020; Zahra et al., 2023).

Survivability is a prevalent subject among Kelantan's women entrepreneurs. With few formal career prospects and societal demands placed on them, many women turn to entrepreneurship to survive. Furthermore, many single mothers must provide for their families, necessitating small business establishments. Nonetheless, these businesses are frequently informal, limiting their growth potential and inhibiting their capacity to contribute substantially to the economy (Baistaman et al., 2022). Therefore, it is critical to tackle these obstacles head-on to develop a healthy entrepreneurial ecosystem among women in Kelantan. Financial literacy programs targeted to their requirements can provide women entrepreneurs with the knowledge and skills they need to manage their finances efficiently (Gallego-Losada et al., 2023). These programs can open doors to long-term growth and expansion by encouraging financial inclusion and offering access to financial resources (Wang et al., 2020).

Women's entrepreneurial potential can be significantly aided through mentorship programs and other efforts that boost women's skills and confidence. The exchange of knowledge and insights between experienced mentors and aspiring entrepreneurs can result in innovative business ideas and improved problem-solving abilities (Lusardi et al., 2021). Business management, marketing, and networking training programs can also give these enterprising women the tools they need to develop their businesses and contribute more to the economy (Zahra et al., 2023). It is equally crucial in building an environment that recognizes the role of women entrepreneurs while still being business-friendly and conducive to doing business.

Women may overcome informal barriers to economic participation if public policy reforms provide access to formal markets, business support services, and opportunities for economic activity (Baistaman et al., 2022; Lusardi et al., 2021). Kelantan can realize its full potential and open up new channels of economic growth by encouraging more women to pursue entrepreneurship. Hence, this chapter dives into the challenges these women entrepreneurs endure due to financial exclusion, as well as several solutions meant to assist them in adapting to the world of digital finance.

2 The Reality of Non-educated Women Entrepreneurs

A significant number of Kelantanese women who own businesses have a limited formal education as a range of socioeconomic factors cause this. As a result of the lack of education, individuals may have restricted access to information, financial resources, and current technology (Zahra et al., 2023). On top of that, as the globe shifts toward digital finance solutions, these women risk being left behind from the opportunities the digital economy provides. The reality of uneducated women entrepreneurs is a significant and frequently disregarded facet of entrepreneurship. While education is primarily acknowledged as a critical aspect in empowering individuals to thrive in business, numerous women entrepreneurs have succeeded despite a lack of formal education. This debate focuses on the challenges and opportunities that non-educated women entrepreneurs have encountered and their dedication and commitment in the face of adversity.

One of the most significant challenges that non-educated women entrepreneurs encounter is a lack of resources and chances (Zainol & Al Mamun, 2018). These women may face barriers to funding, mentorship, and business training programs that their educated peers can access easily. Thus, due to this lack of support, it may be more difficult for individuals to launch and develop their businesses, putting them at a disadvantage in the highly competitive environment of the business community. In addition, women entrepreneurs without a formal education may have to contend with the prejudices and biases of society, both of which work to damage their credibility and ability. There is a widespread misconception that education is required for business accomplishment, which leads to skepticism or doubt in the talents of non-educated entrepreneurs (Zahra et al., 2023; Zainol & Al Mamun, 2018). Overcoming these preconceived views can be emotionally exhausting, and it may necessitate more effort for these women to establish themselves and acquire the trust of investors, consumers, and stakeholders.

On the other hand, despite these obstacles, non-educated women entrepreneurs have demonstrated impressive perseverance and resourcefulness in growing their businesses. Many rely on their practical talents, intuition, and life experiences to successfully navigate the business environment (Zainol & Al Mamun, 2018). Their desire to better their lives and support their families motivates them to seek alternate paths to entrepreneurship that capitalize on their natural talents and passions. Furthermore, non-educated women entrepreneurs frequently bring a distinct viewpoint to their businesses, as their knowledge and experiences can give novel solutions to real-world challenges. Their capacity to form personal connections with customers can be a great advantage, assisting in fostering both loyalty and trust among those who do business with them (Gallego-Losada et al., 2023).

Hence, to effectively assist aspiring businesswomen lacking formal education, one must take an integrated strategy considering these women's distinctive obstacles (Zainol & Al Mamun, 2018). They may be able to learn the necessary skills and build their businesses if they have access to mentorship programs, business training, and microfinancing options that are targeted to their specific needs. In addition, by combating social biases and working to increase diversity within the entrepreneurship ecosystem, we can make it easier for these women to thrive in an atmosphere conducive to their success.

In conclusion, the reality of non-educated women entrepreneurs sheds light on the resiliency and perseverance of these individuals who pursue success in the business sector despite the tremendous obstacles they face. Unlocking their potential and contributing to a more diverse and inclusive landscape for entrepreneurs can be accomplished through recognizing and promoting these individuals' distinctive abilities and making available specific resources (Laffineur et al., 2018). We can build an environment that embraces varied perspectives and celebrates the spirit of entrepreneurship in all its manifestations by appreciating the contributions of non-educated women entrepreneurs.

3 Digital Divide and Limited Technological Exposure

Non-educated women in Kelantan face a challenging environment due to the digital divide. Many are unfamiliar with new technology improvements, making it difficult for them to navigate digital financial services efficiently. This disparity extends the gap between those who can benefit from digital finance and those who are left behind, limiting the economic growth prospects of these women (Ozili, 2020). Therefore, measures must be taken to eradicate the digital divide and limit technological exposure among women entrepreneurs in Kelantan to support these women and encourage them to grow their businesses (Sultana et al., 2015). Besides that, women entrepreneurs may be unable to take full advantage of the digital age due to differences in access to and use of technology between men and women (Zainol & Al Mamun, 2018). This debate looks at some potential ways to close the digital divide and increase women business owners' exposure to technology in Kelantan:

1. Digital Literacy Programs: One efficient strategy to improve women entrepreneurs' technology proficiency and self-assurance is implementing specific digital literacy programs designed to meet their needs. These programs ought to instruct participants in fundamental computer skills, Internet usage, social media marketing, and the management of online businesses (Ozili, 2020). It is possible to enable women entrepreneurs to leverage the power of digital tools for their businesses by providing them with hands-on training and mentoring opportunities.
2. Access to Digital Infrastructure: Ensuring that women entrepreneurs in Kelantan can access affordable and dependable Internet connectivity. Like many other

rural and neglected areas, Kelantan needs investment from governments and essential stakeholders to build its digital infrastructure. Collaborations between government organizations, telecommunications companies, and community-based organizations may benefit in this endeavor (Pazarbasioglu et al., 2020).

3. Mobile Technology and Apps: Utilizing mobile technology may eliminate the digital divide, as many women in Kelantan might impose access to smartphones despite not having desktops or laptops. Women entrepreneurs can conduct business transactions, access markets, and network effectively by developing mobile applications and platforms that cater to their particular demands (Gallego-Losada et al., 2023).
4. Digital Inclusion in Entrepreneurship Programs: Integrating digital skills training into entrepreneurship programs and seminars may give women entrepreneurs the capabilities to use technology to expand their businesses. Governmental bodies, nonprofits, and universities can collaborate to provide courses that teach women entrepreneurs all they need to know about running a business in the digital era (Lusardi et al., 2021).
5. Mentorship and Support Networks: Women entrepreneurs in Kelantan would benefit significantly from increased access to technology if they had access to mentors and other forms of support. Mentorship programs can help women manage the digital world and beat any obstacles they may face by giving them advice and support (Zainol & Al Mamun, 2018).
6. Promoting Digital Marketing and E-commerce: Women entrepreneurs can reach more people outside their local markets if encouraged to use digital marketing and e-commerce (Pazarbasioglu et al., 2020). They can be equipped to reach a wider audience by attending workshops and seminars on utilizing social media platforms and online marketplaces.
7. Advocacy for Gender-Inclusive Policies: It is critical to advocate for gender-inclusive policies that address the digital divide and encourage women entrepreneurs. Governments and policymakers must prioritize ensuring equitable access to technology and support programs that encourage their engagement in the digital economy (Laffineur et al., 2018).
8. Collaboration with Tech Companies: Working with technology firms may provide Kelantanese women entrepreneurs access to tools and resources to help their businesses thrive (Zahra et al., 2023). Tech businesses might provide discounted or free software solutions and digital services to women entrepreneurs to use the financial technology solution efficiently.

To summarize, tackling the digital divide and insufficient technical exposure among Kelantan's women entrepreneurs necessitates a comprehensive approach. Women entrepreneurs may leverage the power of technology to build their businesses while contributing to the local economy by offering digital literacy training, assuring access to digital infrastructure, using mobile technologies, and establishing mentoring and support networks (Lusardi et al., 2021). By providing training in digital literacy, ensuring access to digital infrastructure, utilizing mobile technologies, and building mentorship and support networks, women entrepreneurs have the potential to capitalize on the power of technology to expand their businesses while also making a positive contribution to the economy of their communities (Baistaman et al., 2022). Therefore, using technology to encourage women entrepreneurs in Kelantan may contribute to more comprehensive economic growth and gender equality.

4 Coping with the Current Digital Finance Era

In order to eradicate the issue of financial exclusion, non-educated women entrepreneurs must be equipped with digital finance knowledge and abilities. Specialized training programs, workshops, and mentoring efforts may be established to introduce them to digital financial tools and improve their confidence in using these technologies to run their businesses successfully. It can be accomplished by providing them with an overview of the tools. For women entrepreneurs in Kelantan, recognizing the benefits and problems presented by the modern digital finance age may be daunting. It is imperative as digital financial technologies revolutionize the business environment (Pazarbasioglu et al., 2020). Therefore, it is crucial to address the particular challenges that women entrepreneurs encounter and give them adequate solutions to survive in this new era.

4.1 Issues in Implementing Digital Finance

1. Limited Access to Digital Financial Services: Digital financial services, such as online banking, mobile payment platforms, and digital loans, may be difficult for women entrepreneurs in Kelantan. Their capacity to effectively manage their money and get loans for the expansion of their businesses may be impeded by limited access.
2. Digital Literacy and Awareness: Many women entrepreneurs may not be very tech-savvy or know much about the digital financial tools available to them. They may be unable to properly utilize digital financial services or benefit from their advantages due to their digital illiteracy.
3. Gender Bias and Discrimination: Access to funding and financing sources for women entrepreneurs could be affected by gender bias and discrimination in the finance sector. Compared to their male counterparts, they may face challenges obtaining loans or investment opportunities.
4. Cybersecurity Concerns: The risks associated with cybercrime, i.e., breach of data, Internet fraud, and identity theft, are exposed to women entrepreneurs by their participation in digital finance. They may be reluctant to embrace digital financial solutions due to the fear of cyber threats.

We like to propose the resolutions for the issues mentioned above in implementing digital finance:

1. Digital Financial Literacy Programs: Providing women entrepreneurs in Kelantan with the requisite skills and competencies for managing a digitized financial world is possible by employing specially formulated digital financial literacy programs (Lusardi et al., 2021). Various topics such as online banking, mobile payment networks, digital invoicing, and security on the Internet should be covered by these programs (Pazarbasioglu et al., 2020).
2. Accessible and User-Friendly Platforms: A user-friendly digital platform can facilitate women entrepreneurs' needs and preferences. Accessibility may be improved by cooperating with financial institutions and fintech companies to develop inclusive platforms (Wang et al., 2020).
3. Gender-Inclusive Financial Policies: A level playing field can be achieved by advocating gender-inclusive financial policies addressing women entrepreneurs' needs. It can include promoting equal access to credit, fair lending practices, and support for women-run businesses (Laffineur et al., 2018).
4. Mentorship and Networking Opportunities: Valuable guidance and insight can be provided by establishing mentoring and networking programs for women entrepreneurs that bring together experienced professionals in digital financial services (Lusardi et al., 2021). Building confidence in using digital financial instruments can be achieved through mentoring.
5. Cybersecurity Education: Providing cybersafety training and equipment to women entrepreneurs can contribute to raising awareness of the risks that may be posed and empowering them to protect their digital assets. There is also information on the safe storage of confidential data, recognizing phishing efforts, and using reliable Internet platforms (Pazarbasioglu et al., 2020).
6. Collaboration with Financial Institutions: To address the unique needs of women entrepreneurs, it is possible to cooperate with banks and financial institutions to create specially tailored financial products and services. The provision of flexible loan options, microcredits, and business support services may also be part of (Laffineur et al., 2018).
7. Building Digital Ecosystems: Growth and opportunities can be fostered by establishing supportive digital ecosystems that unite women entrepreneurs with strategic partners such as investors, mentors, and customers (Zahra et al., 2023).

In conclusion, there is an urgent need for a whole approach to address particular problems of women's entrepreneurship within Kelantan simultaneously as it responds to today's digital finance era. Women entrepreneurs can confidently enter the digital financial age and unlock its full potential for their businesses by providing digital financial literacy, ensuring better access to digital finance services, supporting policies for equal opportunities between women and men, and increasing cybersecurity awareness.

4.2 Building Financial Literacy

One key aspect of supporting women entrepreneurs in Kelantan is to improve their ability to learn about finance. To enable them to make a proper decision concerning the financing of their business and improve its overall soundness, these women can be provided with adequate training in finance, including budgeting, accounting, and payment methods by electronic means (Laffineur et al., 2018). They must be educated and given the necessary knowledge and skills to make economic business decisions. Here are some strategies for the promotion of financial literacy among women entrepreneurs in the region:

1. Financial Literacy Workshops and Training: Develop specific workshops and seminars tailored to the needs of female entrepreneurs. The topics covered in these sessions are budgeting, cash flow management, finance planning, and understanding financial statements. These workshops should be conducted by finance experts and practitioners who will advise on best practices.
2. Digital Financial Literacy: Educate business women on digital financial tools and platforms which can assist them in controlling their finances more effectively. Such activities include e-banking, digital payment schemes, invoicing applications, and accounting software. Give them hands-on training to make the best use of these digital tools.
3. Peer Learning and Mentorship: Facilitate peer learning groups or mentorship programs that allow experienced women entrepreneurs to share financial knowledge and experience with those just starting. An environment conducive to exchanging knowledge and developing skills can be created through peer learning.
4. Financial Counselling Services: Provide women entrepreneurs in Kelantan with free or subsidized financial counselling. Professional financial advisors can provide individualized support based on these women's financial circumstances and business requirements.
5. Financial Literacy Materials in Local Languages: In order to facilitate access for women entrepreneurs and improve their understanding, provide information on financial entrepreneurship in the local language such as handbooks, brochures, or Internet resources.
6. Collaborate with Financial Institutions: Collaborate with local banks and financial institutions to develop financial literacy programs for women entrepreneurs. In particular, these organizations may provide information on banking services, financing, and credit management.
7. Use Real-Life Case Studies: Bring together the case studies of successful female entrepreneurs who have effectively managed their finances. As a practical example for others to learn from, these case studies can serve as examples.
8. Government Support: Work with public authorities and initiatives that promote financial literacy and entrepreneurship. To implement financial literacy programs for women business operators in Kelantan, seek funding and support.

9. Business Incubators and Accelerators: Work with accelerators and incubators focusing on supporting women entrepreneurs and businesses. There are often resources and guidance on those platforms in finance management.
10. Evaluation and Continuous Improvement: Regularly evaluate how financial literacy programs benefit women entrepreneurs, and take action based on the results.

To summarize, stakeholders can create an enabling environment in Kelantan to encourage women entrepreneurs to become financially literate by implementing these strategies. It will benefit individual women entrepreneurs and contribute to overall economic growth and equality between men and women in the region by empowering them with financial knowledge and skills (Gallego-Losada et al., 2023).

4.3 Simplifying Digital Finance Tools

Digital financial platforms should be user-friendly and straightforward to meet the needs of women entrepreneurs who do not have an education. Accessibility and adoption for this demography can be enhanced through simplified interfaces, the choice of different languages, or the integration of local context. Besides that, to promote financial inclusion and expand access for unbanked populations, simplifying digital finance tools is an essential step (Lusardi et al., 2021). Digital financial tools use technology to offer practical and inexpensive financial solutions, making them particularly significant for people with little access to traditional banking services (Pazarbasioglu et al., 2020). By simplifying these instruments, we can bridge gaps between financial institutions and the unbanked or underbanked to enable them to participate in the formal financial system (Laffineur et al., 2018) easily. Here are some crucial points that need to be considered to discuss the simplification of digital finance tools for increased financial inclusion:

1. User-Friendly Interfaces: It is essential to design user-friendly and simple interfaces so that individuals with differing levels of digital literacy can avail themselves of the tools for financial transactions. Regardless of prior expertise with technology, simple navigation and clear directions help users feel more confident and comfortable using these tools.
2. Local Language and Context: Create digital finance solutions specific to the target population's regional language and cultural setting. It will make it easier for users who may be more comfortable speaking in their native tongue to understand and trust each other.
3. Reduced Complexity: To lower entry barriers, simplify digital finance tools' sign-up and registration procedures. In this way, it would be possible to reduce the number of documentation requirements and simplify verification procedures so that individuals can open accounts and use services more efficiently.

4. Affordable and Transparent Fees: Ensure that users can access the costs of digital finance instruments transparently and affordably. Higher fees can make uptake more difficult.
5. Customizable Features: Users can customize their digital finance tools to consider their particular needs and preferences. By providing flexibility in account features and transaction possibilities, people can customize the tools to meet their financial needs.
6. Education and Training: Provide user training and educational resources to acquire the knowledge and skills to use digital financial instruments efficiently. It can include a tutorial, video, or interactive seminar to help navigate and maximize the benefits of such tools.
7. Customer Support: Provide several easily accessible customer assistance tools, such as hotlines or chatbots, to answer any question or problem users may encounter while using digital financial instruments.
8. Collaboration with Local Communities: To better understand their needs and preferences, we must work with local communities and organizations. Collaboration efforts may lead to more effective and meaningful solutions for digital finance.
9. Security and Privacy: To build user trust, focus on data security and privacy. Protect users' personal information and finance from threats by implementing stringent security measures.
10. Financial Education Campaigns: In order to raise awareness about the benefits of digital financial instruments and dispel myths or misconceptions related to digital financial services, financial education campaigns may be organized.

By making digital finance tools easier and less complicated, we can remove hurdles to financial inclusion and give people the power to take charge of their finances. Consequently, more people can avail themselves of essential financial services, build their resilience, and play an important part in broader economic development and prosperity (Laffineur et al., 2018).

4.4 Collaborative Efforts and Support

Cooperation efforts and support from a wide range of stakeholders, e.g., government, financial institutions, NGOs, and communities, are necessary to address the financing gap. These agencies can contribute to creating a more inclusive financial ecosystem that allows underbanked populations to use essential financial services and resources. Some core collaboration efforts and supporting mechanisms that can help to combat financial exclusion are listed below:

1. Government Initiatives: Governments promote financial inclusion through policies and legislation. They can develop and apply comprehensive financing policies that encourage financial institutions to serve marginalized populations.

Moreover, the provision of financial literacy programs and support for initiatives aimed at disadvantaged groups can be provided by governments.

2. Financial Institutions' Engagement: Financial institutions can support financial inclusion by creating customized products and services for customers with low or no access to finance. Access to banking services for marginalized populations may be improved by providing low-cost accounts, microfinance, and mobile banking solutions.
3. NGO Partnerships: NGOs often have deep-rooted contacts in the community and are familiar with the unique needs of disadvantaged populations. Financial institutions that work with NGOs can make it easier to reach underserved communities efficiently and develop products to solve their specific financial challenges.
4. Technology Innovations: Financial inclusion can be substantially increased using technology, e.g., mobile banking and electronic payment solutions. Innovation may be driven by cooperation between financial technology companies and traditional financial institutions to achieve access to financial services.
5. Financial Education and Literacy Programs: It is necessary to undertake joint efforts to promote financial education and literacy to make informed financial choices. Partnerships between financing institutions, governments, and NGOs can achieve effective finance education initiatives.
6. Public-Private Partnerships: To tackle financial exclusion comprehensively, public and private partnerships can be formed by exploiting the expertise and resources of both sectors. This type of partnership may facilitate the development of integrated financial products and services.
7. Data Sharing and Analytics: Better assessment of the risks and credit scores for underbanked populations can be achieved by working with banks and credit bureau. In case of financial exclusion, access to accurate credit data may improve the availability of loans.
8. Community Outreach: Engaging with communities directly and understanding their needs are essential to address the financing gap. In order to meet specific needs, cooperation efforts in conducting outreach programs and focused discussions may contribute to the development of funding solutions.
9. Advocacy and Awareness Campaigns: The importance of financial inclusion and its benefits to individuals and the economic sector can be promoted in joint efforts by several stakeholders. An advocate campaign can also address cultural or social obstacles preventing financial inclusion.
10. Inclusive Policy Frameworks: To achieve shared goals such as promoting financial inclusion, governments, financial institutions, and stakeholders must cooperate in establishing coherent policy frameworks that will foster cooperation and promote financial inclusion.

Financial exclusion can be effectively addressed by fostering collaborative efforts and support, creating a more equitable and inclusive financial system that benefits all individuals regardless of socioeconomic origin (Gallego-Losada et al., 2023). These collective efforts can enhance economic empowerment, improve financial security, and significantly contribute to sustainable growth and development.

4.5 Financial Inclusion

Financial inclusion provides a comprehensive range of financial services and products to all segments of society, particularly those historically excluded from the formal financial system (Lusardi et al., 2021). It seeks to ensure that individuals and businesses have access to banking services, credit, insurance, and other financial instruments that can assist them in managing their finances, accumulating assets, and participating in economic activities. Financial inclusion is essential for fostering economic growth, reducing poverty, and promoting financial stability on a global scale (Pazarbasioglu et al., 2020).

In contrast, financial exclusion occurs when individuals or groups lack access to conventional financial services and products. Many factors, such as low income, lack of financial literacy, geographical barriers, social discrimination, and restrictive financial regulations, can cause this exclusion. At both the individual and macroeconomic levels, financial exclusion can have substantial negative consequences on the economy (Lusardi et al., 2021).

4.5.1 Impact of Financial Exclusion on the Economy

1. Reduced Economic Growth: If the financial exclusion of a large proportion of the population limits their access to economic activity or investment in businesses, likely, they are adversely affected. Potential entrepreneurs and small enterprises may find it difficult to obtain capital and loans, which could have a negative impact on growth and productivity.
2. Increased Income Inequality: Financial exclusion can worsen income inequality since those not part of the official financial system are frequently denied opportunities for economic progress and asset creation. It can result in a concentration of wealth and resources in the hands of some fortunate individuals.
3. Limited Access to Credit: Individuals and enterprises that lack access to credit may be unable to invest in education, housing, and productive assets. It can also result in reliance on informal lenders or loan sharks, who frequently charge exorbitant interest rates and trap consumers in a cycle of debt.
4. Vulnerability to Economic Shocks: Financially excluded individuals are more susceptible to economic disruptions, such as job loss or illness. They may struggle to manage unanticipated costs and risks without access to financial services such as insurance and investments.
5. Underutilization of Human Capital: Financial exclusion can make it challenging for competent individuals to leverage their human capital because they cannot

access sufficient funds to spend on their education and skills. It can slow the growth of human capital as a whole and make the economy less productive.

6. Informal Economy Expansion: Individuals may turn to informal financial mechanisms such as money lenders or rotating savings and credit associations (ROSCAs) in the absence of formal financial services. While these mechanisms can help, they frequently lack consumer protections and contribute to the rise of the informal economy.

In order to achieve a level playing field and sustained economic growth, the problem of financing exclusion must be addressed. In financial inclusion, by employing measures such as building a legal environment that facilitates access to accessible financial services, increasing awareness about financial literacy, and using technology to reach out to unreached populations, governments, finance institutions, and policymakers can make a significant contribution. A holistic approach to addressing the challenges faced by Kelantan women entrepreneurs is required to promote financial inclusion and provide them with support for their successful integration into the system. Some strategies for helping women entrepreneurs in Kelantan get access to finance are given below:

1. Financial Literacy Programs: Introduce tailor-made programs to make women entrepreneurs financially literate in Kelantan. These programs should focus on essential management of finances, budgets, access to credit, and knowledge of financial products. Make them aware of the knowledge and skills needed to ensure the efficient functioning of the financial system.
2. Access to Financial Services: Expand the formal financial services available to Kelantan women's businesses. It can be done by setting up financial kiosks, mobile banking units, or joint ventures with nearby banks to provide specialized goods for women entrepreneurs.
3. Microfinance and Credit: Provide women's business owners access to microfinance and credit. To mitigate the risks financial institutions perceive when lending to women entrepreneurs, they should set up microfinance institutions or provide loan guarantees.
4. Women Business Incubators: Set up women's business incubators to mentor, network, and finance the development of new businesses. With these incubators, it is possible to foster the growth of women's businesses and create an entrepreneurial ecosystem.
5. Digital Financial Services: Support women business owners in adopting digital financial services. To increase the convenience and efficiency of commercial transactions, promote mobile banking technologies, digital payment systems, or online platforms for business.
6. Access to Markets: Help women entrepreneurs access markets and supply chains. They organize trade fairs, business networking events, and workshops to establish contacts with interested buyers and customers.
7. Mentoring and Networking: Develop mentorships and networking programs connecting experienced women entrepreneurs with those seeking to begin their

careers. Peer support and advice can be useful in dealing with problems and building confidence.

8. Capacity Building and Skills Development: In areas such as sales, product development, and technology adoption, offer training programs and workshops to increase women entrepreneurs' competencies and capacities.
9. Government Support and Incentives: Advocate Kelantan for government support and incentives targeting female entrepreneurs. Tax incentives, grants, and procurement opportunities may form part of this.
10. Collaborations and Partnerships: In order to achieve a collective solution of the needs of women entrepreneurs, collaboration and partnerships between public authorities, NGOs, financial institutions, and private sector companies are needed.
11. Empowerment Through Education: To improve their leadership and management abilities, support women's businesses in pursuing higher education and professional development.
12. Promote Awareness and Change Mindsets: The importance of gender equality and women's contribution to the economy needs to be highlighted, to combat gender stereotypes and cultural norms that can prevent women from playing their whole part in the economy.

By implementing these strategies and adhering to a gender-inclusive strategy, Kelantan can foster an atmosphere where female entrepreneurs may prosper, fostering the state's social and economic development. Economic empowerment for women benefits them as individuals, their families, the communities, and the nation's economy.

5 Conclusion

In conclusion, it is vital to encourage inclusive growth and reduce financial exclusion by empowering women entrepreneurs from all walks of life at Kelantan who cannot compete in the digital finance era. These women can benefit from the digital economy and play a significant role in state economic development by helping them acquire tailored training on digital finance, enhancing financial literacy, and simplifying digital finance tools. In addition, creating a conducive environment that would boost and support these women's journey toward financial inclusion and entrepreneurial success is crucial to stakeholder cooperation. Malaysia can bring about a more inclusive and equal economic landscape for all its citizens through these common actions.

In Kelantan's families and communities, women entrepreneurs play a significant role through which they have often provided their main source of income. In order to unlock their potential and enable them to achieve sustainable growth, there is a need to overcome the challenges they face, like a lack of knowledge about finance and informal business practices. Thus, by fostering their entrepreneurial spirit through mentorship, training, and policy support, Kelantan can pave the way for a flourishing, inclusive, and mutually beneficial entrepreneurial ecosystem.

References

Baistaman, J., Nawi, F. A. M., Mustapha, W. M. W., & Mamat, M. (2022, September). Exploratory factor analysis of financial literacy in the Malaysian context. In *Proceedings* (Vol. 82, No. 1, p. 24). MDPI. doi:https://doi.org/10.3390/proceedings2022082024.

Gallego-Losada, M. J., Montero-Navarro, A., García-Abajo, E., & Gallego-Losada, R. (2023). Digital financial inclusion. Visualizing the academic literature. *Research in International Business and Finance, 64*, 101862.

Laffineur, C., Tavakoli, M., Fayolle, A., Amara, N., & Carco, M. (2018). Insights from female entrepreneurs in MENA countries: Barriers and success factors. *Entrepreneurship Ecosystem in the Middle East and North Africa (MENA) Dynamics in Trends, Policy, and Business Environment*, pp. 351–397.

Lusardi, A., Hasler, A., & Yakoboski, P. J. (2021). Building up financial literacy and financial resilience. *Mind and Society, 20*, 181–187.

Ozili, P. K. (2020). Theories of financial inclusion. In E. Özen & S. Grima (Eds.), *Uncertainty and challenges in contemporary economic behaviour (Emerald studies in finance, insurance, and risk management)* (pp. 89–115). Emerald Publishing Limited.

Pazarbasioglu, C., Mora, A. G., Uttamchandani, M., Natarajan, H., Feyen, E., & Saal, M. (2020). *Digital financial services*. World Bank, 54.

Sultana, A. M., Arifin, N. S., & Juraini, J. O. (2015). Women involvement in small scale business in Kota Bharu, Kelantan, Malaysia. In *Kertas kerja dibentangkan di 8th Asia-Pacific Business Research Conference, Hotel Istana, Kuala Lumpur pada* (Vol. 9).

Wang, Q., Yang, J., Chiu, Y. H., & Lin, T. Y. (2020). The impact of digital finance on financial efficiency. *Managerial and Decision Economics, 41*(7), 1225–1236.

Zahra, S. A., Liu, W., & Si, S. (2023). How digital technology promotes entrepreneurship in ecosystems. *Technovation, 119*, 102457.

Zainol, N. R., & Al Mamun, A. (2018). Entrepreneurial competency, competitive advantage and performance of informal women micro-entrepreneurs in Kelantan, Malaysia. *Journal of Enterprising Communities: People and Places in the Global Economy, 12*(3), 299–321.

N. Fadzlyn is presently immersed in her pursuit of knowledge as a Ph.D student at Taylor's University, Malaysia. Her academic journey has been marked by diverse experiences, starting a corporate career as an Assistant Company Secretary in one of the companies in Kuala Lumpur and a Secretarial Trainee at Sime Darby Berhad. Academically, she hold a Master's degree in Corporate Administration with graduates ICSA, Bachelor of Corporate Administration majoring in Corporate Governance and Foundation in Law from Universiti Teknologi Mara (UiTM) Malaysia. Professionally, she is recognised as a member of the Malaysian Institute of Chartered Secretaries and Administrators (MAICSA). Beyond the academic and professional realms, she is also one of the women entreprenuers in Kelantan who has her own cosmetic line, named Corlyn Beaute.

Nik Maheran Nik Muhammad is currently a Professor at Universiti Malaysia Kelantan (UMK). Prior to joining UMK, she started her career as an Investment analyst at Kuala Lumpur Stock Exchange (KLSE) and Inland Revenue Board of Malaysia (IRB). Academically, she holds a Doctorate in Business Administration specialized in Investment Decision Making from the University Sains Malaysia, Master's of Business Administration and Bachelor of Business Administration majoring in Finance from the University of Wisconsin Milwaukee, USA. She was also sitting for postdoctoral position and a visiting honorary at the University of Melbourne, Australia. Professionally, she is a CITM (chartered Institute Taxation of Malaysia), FPAM (Financial Planning Association of Malaysia), Institute of Bankers Malaysia (IBBM), Alumni for International volunteer leadership Program (IVLP) of United States and a Distinguish Fellow of Global Strategic Management Inc. USA. Apart from teaching and research, she is also a distinguish speaker, trainer, and consultant in the area of financial behavior, financial psychology, investment decision, entrepreneurship, and professional image. She is a co-founder of Beehive Malaysia and a founder of Women Entrepreneurship Institute (WEC), a project under United States of Government for Social Innovation and Entrepreneurship for youth and women.

MIX
Papier aus verantwortungsvollen Quellen
Paper from responsible sources
FSC® C105338

If you have any concerns about our products,
you can contact us on
ProductSafety@springernature.com

In case Publisher is established outside the EU,
the EU authorized representative is:
Springer Nature Customer Service Center GmbH
Europaplatz 3, 69115 Heidelberg, Germany

Printed by Libri Plureos GmbH
in Hamburg, Germany